Applied Mathematics for Physical Chemistry

THIRD EDITION

JAMES R. BARRANTE

Department of Chemistry
Southern Connecticut State University

Upper Saddle River, NJ 07458

Library of Congress Cataloging-in-Publication Data
Barrante, James R.
Applied mathematics for physical chemistry / James R. Barrante.-- 3rd ed.
p. cm.
Includes bibliographical references and index.
ISBN 0-13-100845-5 (pbk.)
1. Chemistry, Physical and theoretical--Mathematics. I. Title.
QD455.3.M3 B37 2004
510'.24'541--dc22

2003015715

Project Manager: *Kristen Kaiser*
Acquisitions Editor: *Nicole Folchetti*
Editor in Chief: *John Challice*
Production Editor: *Donna Young*
Assistant Managing Editor: *Beth Sweeten*
Executive Managing Editor: *Kathleen Schiaparelli*
Marketing Manager: *Steve Sartori*
Manufacturing Buyer: *Alan Fischer*
Manufacturing Manager: *Trudy Pisciotti*
Copy Editor: *Write With, Inc.*
Proofreader: *Sandra J. Price*
Art Director: *Jayne Conte*
Cover Designer: *Stephen Barrante*

Pearson Prentice Hall
Pearson Education, Inc.
Upper Saddle River, NJ 07458

Printed in the United States of America

ISBN 0-13-100845-5

Pearson Education LTD., *London*
Pearson Education Australia PTY, Limited, *Sydney*
Pearson Education Singapore, Pte. Ltd
Pearson Education North Asia Ltd, *Hong Kong*
Pearson Education Canada, Ltd., *Toronto*
Pearson Educación de Mexico, S.A. de C.V.
Pearson Education—Japan, *Tokyo*
Pearson Education Malaysia, Pte. Ltd

Contents

Preface vii

1 Coordinate Systems 1

1-1 Introduction 1
1-2 Cartesian Coordinates 1
1-3 Plane Polar Coordinates 3
1-4 Spherical Polar Coordinates 5
1-5 Complex Numbers 7

2 Functions and Graphs 11

2-1 Functions 11
2-2 Graphical Representation of Functions 12
2-3 Roots of Polynomial Equations 21

3 Logarithms 24

3-1 Introduction 24
3-2 General Properties of Logarithms 24

3-3 Common Logarithms 25
3-4 Natural Logarithms 27

4 Differential Calculus 30

4-1 Introduction 30
4-2 Functions of Single Variables 31
4-3 Functions of Several Variables; Partial Derivatives 36
4-4 The Total Differential 39
4-5 Derivative as a Ratio of Infinitesimally Small Changes 41
4-6 Geometric Properties of Derivatives 45
4-7 Constrained Maxima and Minima 49

5 Integral Calculus 57

5-1 Introduction 57
5-2 Integral as an Antiderivative 57
5-3 General Methods of Integration 58
5-4 Special Methods of Integration 60
5-5 The Integral as a Summation of Infinitesimally Small Elements 64
5-6 Line Integrals 67
5-7 Double and Triple Integrals 70

6 Infinite Series 76

6-1 Introduction 76
6-2 Tests for Convergence and Divergence 77
6-3 Power Series Revisited 81
6-4 Maclaurin and Taylor Series 83
6-5 Fourier Series and Fourier Transforms 85

7 Differential Equations 95

7-1 Introduction 95
7-2 Linear Combinations 96
7-3 First-Order Differential Equations 97
7-4 Second-Order Differential Equations with Constant Coefficients 100
7-5 General Series Methods of Solution 104
7-6 Special Polynomial Solutions of Differential Equations 106
7-7 Exact and Inexact Differentials 113
7-8 Integrating Factors 116
7-9 Partial Differential Equations 117

8 Scalars and Vectors 123

8-1 Introduction 123
8-2 Addition of Vectors 123
8-3 Multiplication of Vectors 126
8-4 Applications 128

9 Matrices and Determinants 132

9-1 Introduction 132
9-2 Square Matrices and Determinants 133
9-3 Matrix Algebra 135
9-4 Solutions of Systems of Linear Equations 139
9-5 Characteristic Equation of a Matrix 141

10 Operators 149

10-1 Introduction 149
10-2 Vector Operators 151
10-3 Eigenvalue Equations Revisited 153
10-4 Hermitian Operators 155
10-5 Rotational Operators 156
10-6 Transformation of ∇^2 to Plane Polar Coordinates 160

11 Numerical Methods and the Use of the Computer 165

11-1 Introduction 165
11-2 Graphical Presentation 169
11-3 Numerical Integration 175
11-4 Roots of Equations 179
11-5 Fourier Transforms Revisited: Macros 182

12 Mathematical Methods in the Laboratory 193

12-1 Introduction 193
12-2 Probability 193
12-3 Experimental Errors 196
12-4 Propagation of Errors 200
12-5 Preparation of Graphs 204
12-6 Linear Regression 205
12-7 Tangents and Areas 207

Appendices 211

I Table of Physical Constants 211
II Table of Integrals 211
III Transformation of ∇^2 to Spherical Polar Coordinates 225
IV Stirling's Approximation 228
V Solving a 3×3 Determinant 229
VI Statistics 230

Answers 232

Index 241

Preface

A perusal of many modern physical chemistry texts demonstrates that most authors of these texts and the professors who use them, such as myself, expect students to know a great deal more mathematics than is covered in the calculus courses normally required for physical chemistry courses. Moreover, we honestly expect that our students will know how to apply the mathematics they have learned to physical problems. Unfortunately, many of my colleagues and I have found that this generally is not the case. It was this observation, along with the fact that I was spending a great deal of lecture time teaching mathematics rather than physical chemistry in my physical chemistry course, that inspired me to write the first edition of this text some 30 years ago.

It is my intention, therefore, that this third edition be used as a supplement along with the student's physical chemistry textbook, to help students either review or, perhaps, learn for the first time those areas of mathematics that are essential to an understanding of physical chemistry, and, more importantly, to apply that mathematics to physical problems. The purpose of the book is *not* to replace the mathematics courses that are prerequisite to the physical chemistry course, but to be a *how to do it* review mathematics textbook. Consequently, the problems at the end of each chapter are designed to test the reader's mathematical skills, not his or her skills in solving physical chemistry problems.

Like the first two editions, the first five chapters concentrate on subject matter normally covered in prerequisite mathematics courses and should be a review. Again, an emphasis in the chapter on integral calculus has been placed on using integral tables, and, in keeping with the original intent of the book, mathematical rigor was kept at a minimum, giving way to intuition where possible.

The latter half of the text covers important material normally not covered in prerequisite courses, but, for the most part, at an introductory level. For example, the chapter on differential equations emphasizes the solution of second-order linear differential equations with constant coefficients, common to many simple problems in wave mechanics. Also, as in the second edition, sections on the series method of solving differential equations are included. The sections on Fourier series and Fourier transforms have been expanded in this edition to include discrete Fourier transforms and well as continuous Fourier transforms. Discrete Fourier transforms are important in many areas of spectroscopy, since they can be handled by digital computers.

Finally, the chapter on numerical methods has been completely revised. In the second edition, we concentrated on writing programs using BASIC to do the numerical calculations. Over the recent years, however, there has been a move away from using compiled programs for doing scientific computations toward the use of spreadsheets, such as Microsoft Excel®, for such computations. Thus, the new chapter concentrates on using a spreadsheet to do many standard numerical calculations, such as numerical integration, fitting curves to experimental data, and finding discrete Fourier transforms of functions.

As I mentioned in the Preface to the second edition, a text such as this could not be a success without the contributions of a number of people. I especially wish to thank Professor John Bopp, Nazareth College; Professor Wayne Bosma, Bradley University; and Professor Greg Peters, University of Memphis for their careful and critical review of the second edition and their many helpful comments and suggestions. I also would like to thank Professor John Wheeler of the University of California, San Diego, for finding a serious error in one of the examples in the chapter on infinite series in the second edition that survived from the first edition.

I thank my editor John Challice, Project Manager Kristen Kaiser, Production Editor Donna Young, and all those individuals at Prentice Hall ESM and Write With, Inc. who helped to improve immensely the quality of the text.

Finally, I wish to thank my son, Stephen Barrante, who designed the cover for this edition, my wife Marlene, and our family for their patience and encouragement during the preparation of this book.

I welcome comments on the text and ask that any comments or errors found be sent to me at jrbarrante@aol.com.

JAMES R. BARRANTE

CHAPTER 1 Coordinate Systems

1-1 INTRODUCTION

A very useful method for describing the functional dependency of various properties of a physicochemical system is to assign to each property a point along one of a set of axes, called a *coordinate system*. The choice of coordinate system used to describe the physical world will depend to a great extent on the nature of the properties being described. For example, human beings are very much "at home" in rectangular space. Look around you and note the number of 90° angles you see. Therefore, architects and furniture designers usually employ rectangular coordinates in their work. Atoms and molecules, however, live in "round" space. Generally, they operate under potential-energy fields that are centrally located and therefore are best described in some form of "round" coordinate system (e.g., plane polar or spherical polar coordinates). In fact, many problems in physical chemistry dealing with atoms or molecules *cannot* be solved in linear coordinates. We shall see, however, that things we are accustomed to studying in linear coordinates, such as waves, look very different to us when they are described in polar coordinates. Once we have come to terms with polar coordinates, however, we will find that they are no more difficult to use than linear coordinates. We begin our discussion with a general treatment of linear coordinates.

1-2 CARTESIAN COORDINATES

In the mid-17th century, the French mathematician René Descartes proposed a simple method of relating pairs of numbers as points on a rectangular plane surface, today called a *rectangular Cartesian coordinate system*. A typical two-dimensional Cartesian coordinate system consists of two perpendicular axes, called the *coordinate axes*. The vertical axis generally is labeled the y-axis, while the horizontal axis generally is labeled the x-axis.[1] The point of intersection between the two axes is called the *origin*.

[1]The correct mathematical designations of the vertical axis and the horizontal axis are the *ordinate* and the *abscissa*, respectively. These terms, however, have fallen out of use over the years.

The application of mathematics to the physical sciences requires taking these abstract collections of numbers and the associated mathematics and relating them to the physical world. Thus, the aforementioned x's and y's could be pressures, volumes, or temperatures describing the behavior of a gas or could be any pair of physical variables related to each other. For example, Fig. 1-1 shows the relationship of the vapor pressure of water to its temperature. Note that, in designating any point on a coordinate system, the x-coordinate is always given first. Thus, the notation (333,149.4) refers to the point on the coordinate system shown in Fig. 1-1 whose T-coordinate is 333 K and whose P-coordinate is 149.4 mm Hg.

The curve depicted in Fig. 1-1 can be described by the equation $P = Ae^{-\Delta H/RT}$, where ΔH is the heat of vaporization of water, R is the gas constant, and A is a constant. The equation describing the curve could just as easily have been written $y = Ae^{-B/x}$; however, not apparent in either equation is the fact that it only approximates the behavior of the vapor pressure of a liquid with temperature, since the implied conditions that the vapor phase behave ideally and that the heat of vaporization be constant with temperature generally are not true. It is a knowledge of facts such as these that continues to distinguish the science of physical chemistry from "pure mathematics."

Coordinate systems are not limited to plane surfaces. We can extend the two-dimensional rectangular coordinate system just described into a three-dimensional coordinate system by constructing a third axis perpendicular to the xy-plane and passing through the origin. In fact, we are not limited to three-dimensional coordinate systems, but can extend the process to as many coordinates as we wish or need—to n-dimensional coordinate systems—although, as we might expect, the graphical representation becomes difficult beyond three dimensions, since we are creatures of three-dimensional space. For example, in wave mechanics, a field of physics having to do with the mechanical behavior of waves, we usually describe the

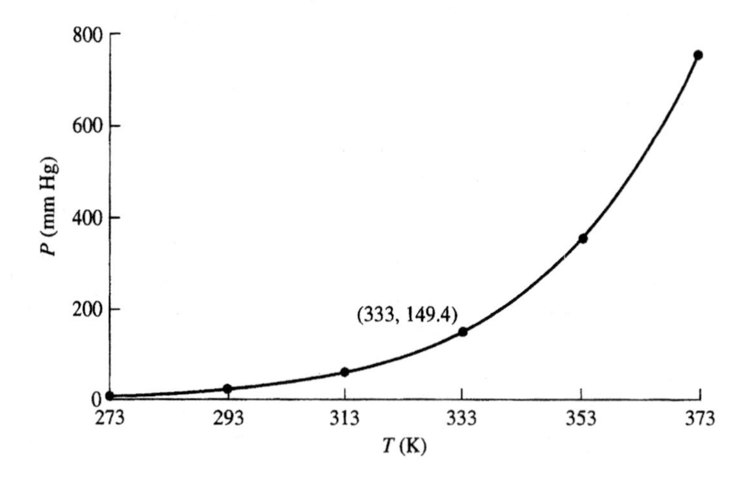

Figure 1-1 Graph of vapor pressure of water versus temperature.

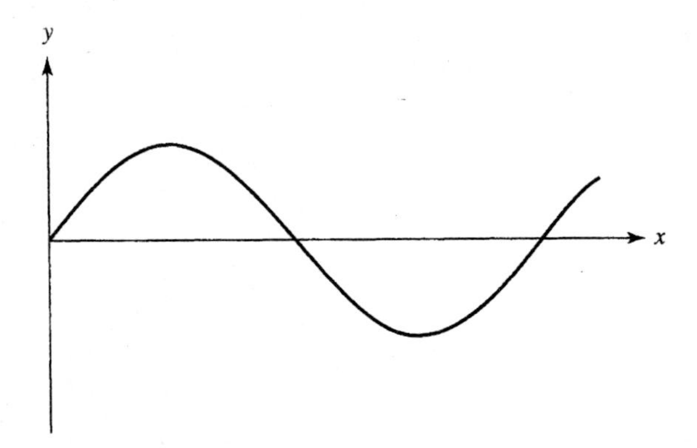

Figure 1-2 Amplitude of a one-dimensional transverse wave.

amplitude of a transverse wave by using an "extra" coordinate. Therefore, the amplitude of a one-dimensional transverse wave—say, along the x-axis—is represented along the y-axis, as illustrated for a simple sine wave in Fig. 1-2. (Keep in mind that the wave itself does not move in the y-direction. Rather, the wave is a disturbance, or a series of disturbances, traveling, in this case, only down the x-axis. We should not confuse the motion of the *wave* with the motion of the *medium producing the wave*.) A two-dimensional transverse wave in the xy-plane, such as a water wave moving across the surface of a lake, requires a third coordinate (call it the z-axis) to describe the amplitude of the wave—still easy to represent graphically and certainly within the realm of our experience. Many problems in wave mechanics, however, require us to describe the amplitudes of three-dimensional transverse waves graphically. In order to do so we need a fourth coordinate. This presents a problem to creatures of three-dimensional space, and a number of ingenious ways have been devised to get around the problem. One way is to use the density of points on a three-dimensional graph to represent the amplitude of the wave.

In any of the coordinate systems described, it is useful to define a very small, or differential, volume element

$$d\tau = dq_1\, dq_2\, dq_3 \ldots dq_n \tag{1-1}$$

where dq_i is an infinitesimally small length along the ith axis. In the three-dimensional Cartesian coordinate system shown in Fig. 1-3, the volume element is simply

$$d\tau = dx\, dy\, dz \tag{1-2}$$

1-3 PLANE POLAR COORDINATES

Many problems in the physical sciences cannot be solved in rectangular space. For this reason, we find it necessary to redefine the Cartesian axes in terms of what

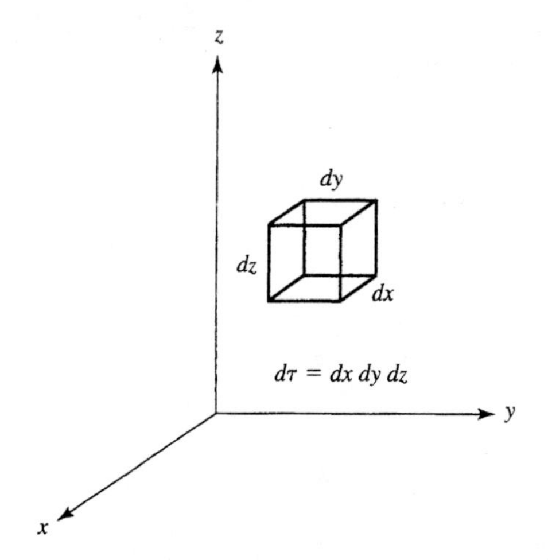

Figure 1-3 Differential volume element for a Cartesian coordinate system.

normally are referred to as "round" or "curvilinear" coordinates. Consider the diagram in Fig. 1-4. It is possible to associate every point on this two-dimensional coordinate system with the geometric properties of a right triangle. Note that the magnitude of x is the same as the length of the side b of the triangle shown and that the magnitude of y is the same as the length of the side a. Since

$$\sin\theta = \frac{a}{r} \quad \text{and} \quad \cos\theta = \frac{b}{r} \tag{1-3}$$

we can write

$$x = b = r\cos\theta \quad \text{and} \quad y = a = r\sin\theta \tag{1-4}$$

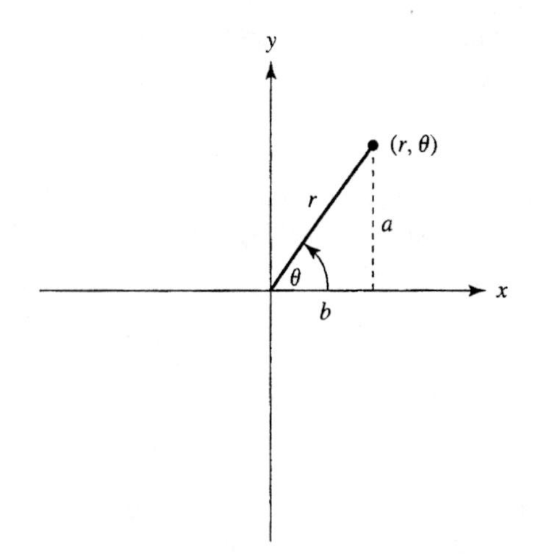

Figure 1-4 Plane polar coordinates.

Therefore, every point (x, y) of the two-dimensional coordinate system can be specified by assigning a value for r and a value for θ at that point. This type of graphical representation is called a *plane polar coordinate system.* In such a coordinate system, points are designated by the notation (r, θ). Equations (1-4) are known as *transformation equations*; they transform the coordinates of a point from polar coordinates to Cartesian (linear) coordinates. The reverse transformation equations can be found by simple trigonometry. We have

$$\frac{y}{x} = \frac{r \sin \theta}{r \cos \theta} = \tan \theta \quad \text{or} \quad \theta = \tan^{-1}\left(\frac{y}{x}\right) \tag{1-5}$$

and

$$r = (x^2 + y^2)^{1/2} \tag{1-6}$$

1-4 SPHERICAL POLAR COORDINATES

While there are not many applications of plane polar coordinates to chemistry, one extension of plane polar coordinates to three dimensions leads to the *spherical polar coordinate system*, shown in Fig. 1-5. This system does have numerous applications to physical chemistry, particularly in those areas associated with the structure of atoms and molecules. A point in the spherical polar coordinate system is represented by three numbers: r, the distance of the point from the origin, normally called the *radius vector*; θ, the angle that the radius vector r makes with the z-axis; and ϕ, the angle that the line $\overline{OA}$ makes with the x-axis. Since

$$\cos \phi = \frac{c}{a}, \quad \sin \phi = \frac{d}{a}, \quad \text{and} \quad \sin \theta = \frac{a}{r} \tag{1-7}$$

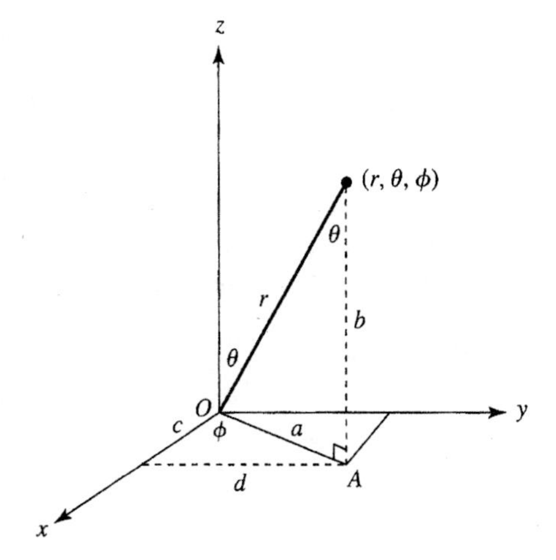

Figure 1-5 Spherical polar coordinates.

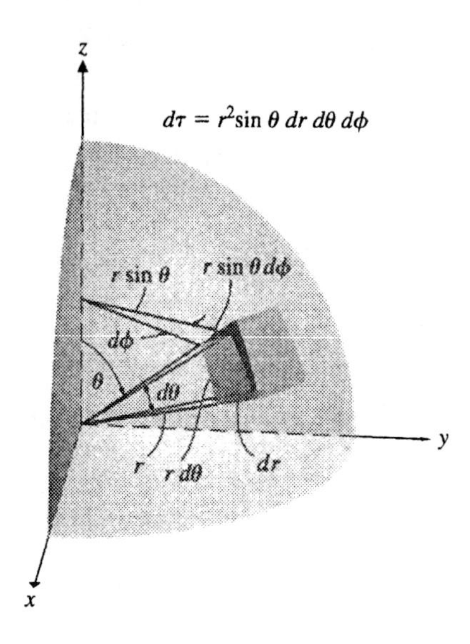

Figure 1-6 Differential volume element in spherical polar coordinates.

and since the length of c is numerically equal to x, d to y, and b to z, we can write

$$\begin{aligned} x &= a \cos\phi = r \sin\theta \cos\phi \\ y &= a \sin\phi = r \sin\theta \sin\phi \\ z &= r \cos\theta \end{aligned} \tag{1-8}$$

The reverse transformation equations are found as follows:

$$\frac{y}{x} = \frac{r \sin\theta \sin\phi}{r \sin\theta \cos\phi} = \tan\phi \quad \text{or} \quad \phi = \tan^{-1}\left(\frac{y}{x}\right) \tag{1-9}$$

$$r = (x^2 + y^2 + z^2)^{1/2} \tag{1-10}$$

$$\cos\theta = \frac{z}{r} \quad \text{or} \quad \theta = \cos^{-1}\frac{z}{(x^2 + y^2 + z^2)^{1/2}} \tag{1-11}$$

The differential volume element in spherical polar coordinates is not as easy to determine as it is in linear coordinates. Recalling, however, that the length of a circular arc intercepted by an angle θ is $L = r\theta$, where r is the radius of the circle, we can see from Fig. 1-6 that the volume element is

$$d\tau = (dr)(r \sin\theta\, d\phi)(r\, d\theta) = r^2 \sin\theta\, dr\, d\theta\, d\phi \tag{1-12}$$

Example

Determine the values of r, θ, and ϕ for the point $(3, -2, 1)$.

Solution

$$x = 3,\ y = -2, \text{ and } z = 1:$$

$$r = \sqrt{x^2 + y^2 + z^2} = \sqrt{9 + 4 + 1} = \sqrt{14} = 3.742$$

$$\theta = \cos^{-1}\left(\frac{z}{r}\right) = \cos^{-1}\left(\frac{1}{3.742}\right) = 74.50°$$

$$\phi = \tan^{-1}\left(\frac{y}{x}\right) = \tan^{-1}\left(\frac{-2}{3}\right) = -33.69°$$

1-5 COMPLEX NUMBERS

A *complex number* is a number composed of a real part x and an imaginary part iy, where $i = \sqrt{-1}$. A complex number is thus represented by the equation $z = x + iy$. Actually, *all* numbers are complex numbers and can be described by the same equation; it is just that for real numbers, $y = 0$, while for pure imaginary numbers, $x = 0$.

Like real numbers, complex numbers can be represented on a coordinate system. The real part of the complex number is designated along the x-axis, while the pure imaginary part is designated along the y-axis, as shown in Fig. 1-7. Since $x = r \cos \theta$ and $y = r \sin \theta$, any complex number can be written as

$$z = x + iy = r \cos \theta + ri \sin \theta = r(\cos \theta + i \sin \theta) \tag{1-13}$$

Moreover, since every point in the plane formed by the x- and y-axes represents a complex number, the plane is called the *complex plane*. In an n-dimensional coordinate system, one plane may be the complex plane.

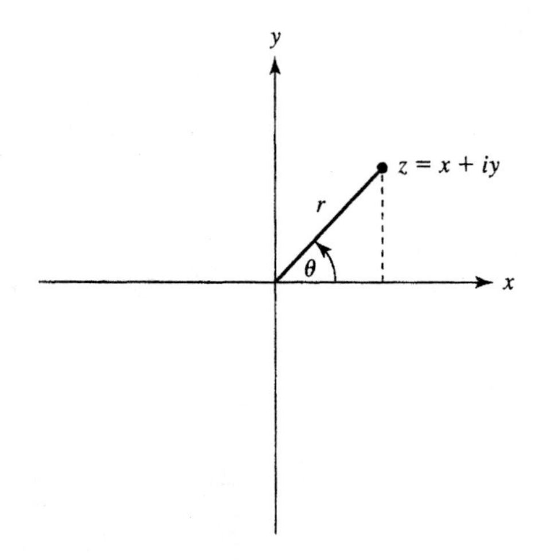

Figure 1-7 The complex plane.

The value of r in Equation (1-13), called the *modulus* or *absolute value*, can be found by the equation

$$r = (x^2 + y^2)^{1/2} = |z| \tag{1-14}$$

The angle θ, called the *phase angle*, simply describes the rotation of z in the complex plane. The square of the absolute value can be shown to be identical to the product of $z = x + iy$ and its complex conjugate $z^* = x - iy$. The complex conjugate of a complex number is formed by changing the sign in front of the imaginary part. The product of a complex number and its complex conjugate is *always* a real number:

$$z^* z = (x - iy)(x + iy) = x^2 + y^2 = |z|^2 \tag{1-15}$$

To find another useful relationship between $\sin\theta$ and $\cos\theta$ in the complex plane, let us expand each function in terms of a Maclaurin series. (Series expansions are covered in detail in Chapter 6.) We have

$$\sin\theta = \theta - \frac{\theta^3}{3!} + \frac{\theta^5}{5!} - \frac{\theta^7}{7!} + - \cdots \tag{1-16}$$

$$\cos\theta = 1 - \frac{\theta^2}{2!} + \frac{\theta^4}{4!} - \frac{\theta^6}{6!} + - \cdots \tag{1-17}$$

where $n!$ (n factorial) is $n! = (1)(2)(3)(4)\ldots(n-1)(n)$. Hence,

$$\cos\theta + i\sin\theta = 1 + i\theta - \frac{\theta^2}{2!} - \frac{i\theta^3}{3!}\cdots \tag{1-18}$$

But this is identical to the series expansion for $e^{i\theta}$:

$$e^{i\theta} = 1 + i\theta - \frac{\theta^2}{2!} - \frac{i\theta^3}{3!}\cdots \tag{1-19}$$

Hence, we can write

$$e^{i\theta} = \cos\theta + i\sin\theta \tag{1-20}$$

and

$$z = re^{i\theta} \tag{1-21}$$

By the same method, it can be shown that

$$e^{-i\theta} = \cos\theta - i\sin\theta \tag{1-22}$$

and

$$z^* = re^{-i\theta} \tag{1-23}$$

Equations (1-20) and (1-22) are known as *Euler*'s (pronounced "oiler's") *relations.*

Example

In the X-ray diffraction of crystals, the structure factor is given by the equation

$$F_{hkl} = \sum_j f_j \, e^{2\pi i(hu_j + kv_j + lw_j)}$$

where h, k, and l are the Miller indices, f_j is the atomic scattering factor of the jth atom, and u_j, v_j, and w_j are the fractional coordinates of the jth atom. Using Euler's relation, transform this equation into an equivalent equation in sines and cosines.

Solution

$$e^{2\pi i(hu_j + kv_j + lw_j)} = \cos[2\pi(hu_j + kv_j + lw_j)] + i \sin[2\pi(hu_j + kv_j + lw_j)]$$

$$F_{hkl} = \sum_j f_j\{\cos[2\pi(hu_j + kv_j + lw_j)] + i \sin[2\pi(hu_j + kv_j + lw_j)]\}$$

SUGGESTED READINGS

1. BRADLEY, GERALD L., and SMITH, KARL J. *Calculus*. Upper Saddle River, NJ: Prentice Hall, 1995.
2. SULLIVAN, MICHAEL. *College Algebra*, 4th ed. Upper Saddle River, NJ: Prentice Hall, 1996.
3. VARBERG, DALE, and PURCELL, EDWIN J. *Calculus*, 7th ed. Upper Saddle River, NJ: Prentice Hall, 1997.
4. WASHINGTON, ALLYN J. *Basic Technical Mathematics*, 6th ed. Boston: Addison-Wesley, 1995.

PROBLEMS

1. What is the sign of the x-coordinate and the y-coordinate of points in each quadrant of a two-dimensional coordinate system? Relate these signs to the sign of $\sin\theta$, $\cos\theta$, and $\tan\theta$ in each quadrant.

2. Determine the values of r and θ for the following points:

(a) $(3, 2)$ **(d)** $(0, 2)$ **(g)** $(-3, 2)$
(b) $(2, \sqrt{2})$ **(e)** $(-2, -2)$ **(h)** $(-5, 0)$
(c) $(1, 4)$ **(f)** $(3, 0)$ **(i)** $(8, -4)$

3. Determine the values of x and y for the following points:

(a) $r = 1.23, \theta = 45.65°$ **(e)** $r = 4.00, \theta = 130°$
(b) $r = 1.00, \theta = 60°$ **(f)** $r = 3.00, \theta = 0°$
(c) $r = \sqrt{3}, \theta = 120°$ **(g)** $r = 2.50, \theta = -35.10°$
(d) $r = 4.65, \theta = 225°$ **(h)** $r = 6.00, \theta = 45°$

4. Determine the values of r, θ, and ϕ for the following points:
   (a) $(1, 2, 3)$ **(c)** $(-2, 0, 1)$ **(e)** $(3, 6, 6)$
   (b) $(1, -2, -3)$ **(d)** $(1, 0, 4)$ **(f)** $(0, 0, 2)$
5. The *cylindrical coordinate system* can be constructed by extending a z-axis from the origin of a plane polar coordinate system perpendicular to the xy-plane. A point in this system is designated (r, θ, z). What is the differential volume element in this coordinate system?
6. Determine the modulus and phase angle for the following complex numbers:
   (a) 1 **(c)** $1 + i$ **(e)** $-3 - 3i$
   (b) $4i$ **(d)** $3 - 4i$ **(f)** $-4 + 6i$
7. Find the roots of the equation $x^3 - 1 = 0$. (Remember that there are three roots and they are not necessarily equal or real.)
8. Show that $e^{-i\theta} = \cos\theta - i\sin\theta$.
9. Show that $\cos\theta = \frac{1}{2}(e^{i\theta} + e^{-i\theta})$ and $\sin\theta = \frac{1}{2i}(e^{i\theta} - e^{-i\theta})$.
10. Find the values of m that satisfy the equation $e^{2\pi im} = 1$. (*Hint*: Express the exponential in terms of sines and cosines, using Euler's relation.)
11. Show that the sum $Ae^{ikx} + Be^{-ikx}$, where A and B are arbitrary constants, is equivalent to the sum $A'\sin kx + B'\cos kx$, where A' and B' are arbitrary constants.
12. In the primitive cubic lattice of CsCl, the Cs ion has fractional coordinates (0, 0, 0) and the Cl ion has fractional coordinates (1/2, 1/2, 1/2). Given that $f_{Cs} = 54$ and $f_{Cl} = 18$, find the X-ray structure factor for the $h = 2, k = 0, l = 0$ reflection, using the equation developed in the example in Section 1-5.

CHAPTER 2

Functions and Graphs

2-1 FUNCTIONS

Physical chemistry, like all the physical sciences, is concerned with the dependence of one or more variables of a system upon other variables of the system. For example, suppose we wish to know how the vapor pressure of a liquid varies with absolute temperature. With a little experimentation in the laboratory, we will find that the vapor pressure of a liquid varies with temperature in a very specific way. Through careful measurements, we will find that the vapor pressure P at any temperature T on the absolute temperature scale can be approximated by the law

$$P = Ae^{-B/T} \tag{2-1}$$

where A and B are experimentally determined constants. This equation predicts that there is a one-to-one correspondence between the vapor pressure of a liquid and its temperature. That is, for every value of T substituted into Equation (2-1), there is a corresponding value of P.

Let us define a collection of temperatures as a mathematical set $T = \{T_1, T_2, T_3, \ldots\}$ and the corresponding pressures as another set $P = \{P_1, P_2, P_3, \ldots\}$. A mathematical set is defined as a collection of numbers, and each member of the set called an *element*, so we see that the collection of temperatures and pressures satisfies this definition. If there is associated with each element of the set T at least one element in the other set P, then this association is said to constitute a function from T to P, written $f{:}T \rightarrow P$. That is, the function takes every element in set T into the corresponding element in set P. Since P is a function of T—that is, the value of P depends on the value of T—the above expression can be written $f{:}T \rightarrow f(T)$, where $P = f(T)$. Remember that $f(T)$, read "f of T," does not mean f multiplied by T; but that $f(T)$ is the value of $Ae^{-B/T}$ at T. Hence, we can write

$$f(T) = Ae^{-B/T} \tag{2-2}$$

A function, then, is defined as a correspondence between elements of at least two mathematical sets.

In the preceding example, P was considered to be a function of only a single variable T. The equation $P = f(T)$ can be represented by a series of points on a two-dimensional Cartesian coordinate system, such as the graph shown in Fig. 1-1. Physicochemical systems, however, usually depend on more than one variable. Thus, it is necessary to extend the definition just given to include functions of more than one variable. For example, the pressure of a gas depends not only on the temperature of the gas, but also on the volume of the gas. If the gas is assumed to be ideal, then the pressure of the gas will follow the well-known ideal gas law

$$P = \frac{RT}{V} = f(T, V) \tag{2-3}$$

where R is a constant. Equation (2-3) implies that there is a one-to-one correspondence between three sets of numbers: a set of pressures $P = \{P_1, P_2, P_3, \ldots\}$, a set of temperatures $T = \{T_1, T_2, T_3, \ldots\}$, and a set of volumes $V = \{V_1, V_2, V_3, \ldots\}$. These three sets can be represented graphically on a three-dimensional coordinate system by plotting P along one axis, T along another axis, and V along the third axis. Such graphs of P, T, and V commonly are called *phase diagrams*, as shown in Fig. 2-1.

2-2 GRAPHICAL REPRESENTATION OF FUNCTIONS

As we have seen, one of the most convenient ways to represent the functional dependence of the variables of a system is with the use of coordinate systems. This is

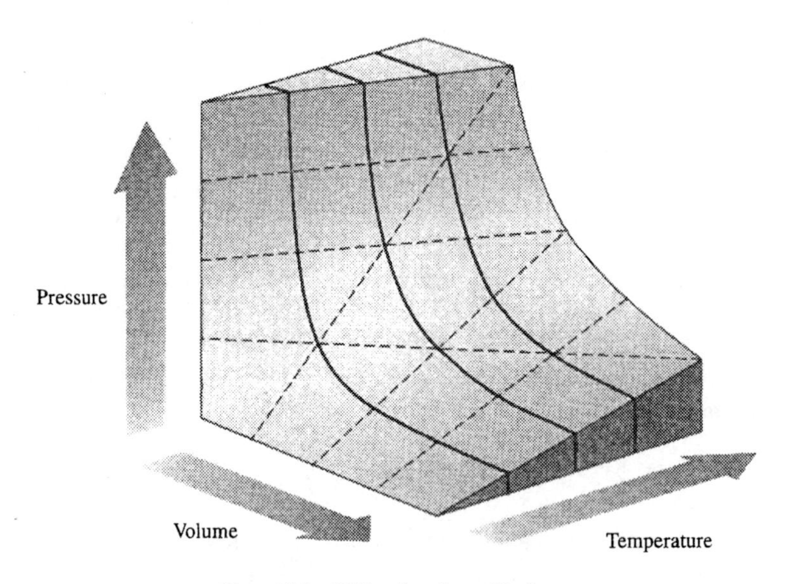

Figure 2-1 *PVT* surface for an ideal gas.

because each set of numbers is easily represented by a coordinate axis, and the graphs that result give an immediate visual representation of their behavior. In this section, we shall explore several types of graphical representation of functions. We begin with functions that describe a linear dependence between variables.

Equations of the First Degree Equations that define functions showing a linear dependence between variables are known as *equations of the first degree* or *first-degree equations.* These functions describe a dependence commonly called the *direct proportion*, and the equations that define them are called first-degree equations because all their variables are raised only to the first power. First-degree equations describe a line and have the general form

$$y = f(x) = mx + b \tag{2-4}$$

where m is a constant called the *slope* of the line and b is a constant called the line's y-*intercept.* The slope m represents the tangent of the angle the line makes with the x-axis, and the y-intercept b represents the point where the line crosses the y-axis. The equation $°F = \frac{9}{5}°C + 32$, illustrated in Fig. 2-2, is a first-degree equation. This familiar equation describes the temperature of a system on the Fahrenheit scale and its relationship to the temperature on the Celsius temperature scale. Note that, indeed, the equation is that of a straight line. From the equation, we see that the slope

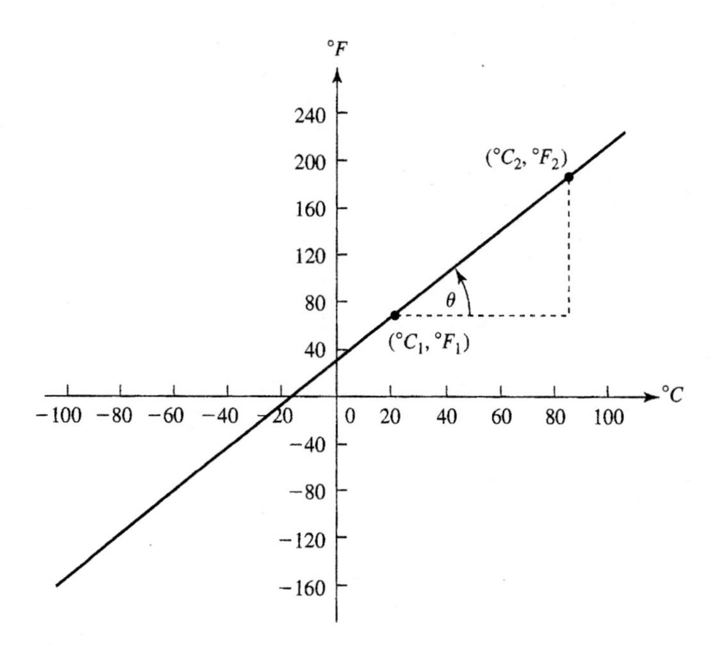

Figure 2-2 Graph of $°F = \frac{9}{5}°C + 32$.

of the line is 9/5 and the y-intercept is 32. It also may be of interest to consider the value of °C for which the value of °$F = f(°C) = 0$. This point represents the point where the line crosses the °C-axis and is known as the *zero of the function*. Rearranging Equation (2-4) gives

$$0 = mx + b \quad \text{or} \quad x = -\frac{b}{m} \tag{2-5}$$

In many cases, equations that do not exhibit a linear relationship between the variables can be made to do so simply by choosing the correct set of coordinate axes on which to plot them. It is useful to do this, because the physical constants in an equation are most easily determined graphically if the equation is linear. For example, the equation representing the vapor pressure of water as a function of temperature, shown in Fig. 1-1, is hardly linear. However, if, instead of plotting P versus T, we plot ln P versus $1/T$, we obtain the graph of $\ln P = -\Delta H/RT + C$, which *is* linear. This graph is shown in Fig. 2-3. The advantage of such a plot is that we can easily determine ΔH, the heat of vaporization of water, from the slope of the line, $-\Delta H/R$. This value would be very difficult to determine from the graph shown in Fig. 1-1.

Equations of the Second Degree Equations of the second degree are those having the general form

$$y = f(x) = ax^2 + bx + c \tag{2-6}$$

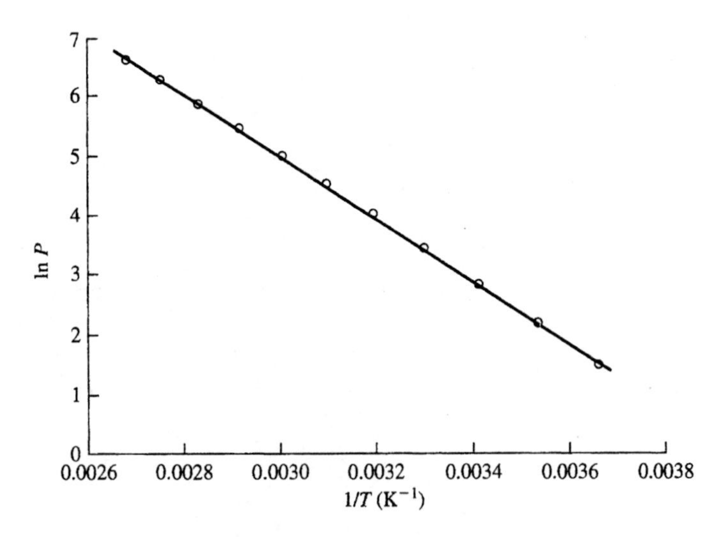

Figure 2-3 Graph of ln P versus $1/T$ for the vapor pressure of water.

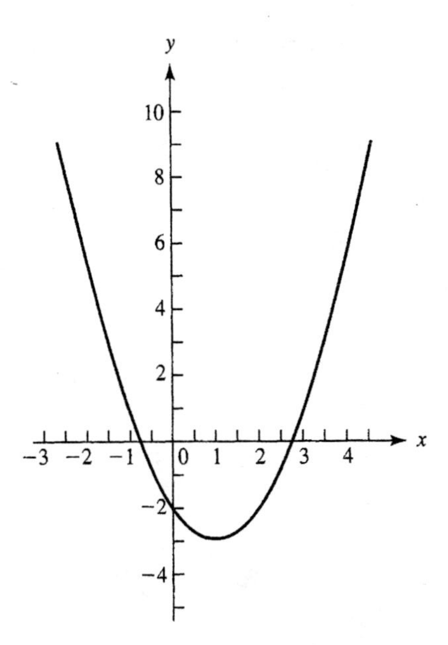

Figure 2-4 Graph of $y = x^2 - 2x - 2$.

where a, b, and c are constants and $a \neq 0$. Thus, a second-degree equation is an equation in which the highest power to which the variable x is raised is 2. Equation (2-6) is given the specific name *quadratic equation*, and the function that it defines is called a *quadratic function*. A typical quadratic equation, $y = x^2 - 2x - 2$, is illustrated in Fig. 2-4. The family of curves that Equation (2-6) describes are called *parabolas*. We can see by experience with this type of curve that when the constant a is positive, the curve opens upward, and when the constant a is negative, the curve opens downward. Notice that the slope of the curve at the point (x, y), which can be defined by a line drawn tangent to the curve at that point, is not constant, but changes as a function of x, as shown in Fig. 2-5. In practice, this type of slope is difficult to determine graphically, since tangent lines to the curve must be constructed. However, we shall see in Chapter 4 that differential calculus can be used to determine the slope of a curve at any point, provided that the equation describing the curve is known.

In the previous section, we defined the zero of the function as the point where a line crosses the x-axis. Equations of higher order than 1 may or may not have more than one zero of the function. We generally refer to the zeros of a function as the *roots* of the associated equation. However, whether the roots of an equation are the zeros of the function will depend on whether the roots are real or imaginary. A second-degree equation must have two roots. If the curve described by the equation crosses the x-axis, then the two roots will be different and will represent the zeros of the function (i.e., the points where $f(x) = 0$). In some cases, the roots of a quadratic equation may be identical. Under those circumstances, the curve described by the equation will not *cross* the x-axis, but will come to zero on the x-axis. Obviously, the

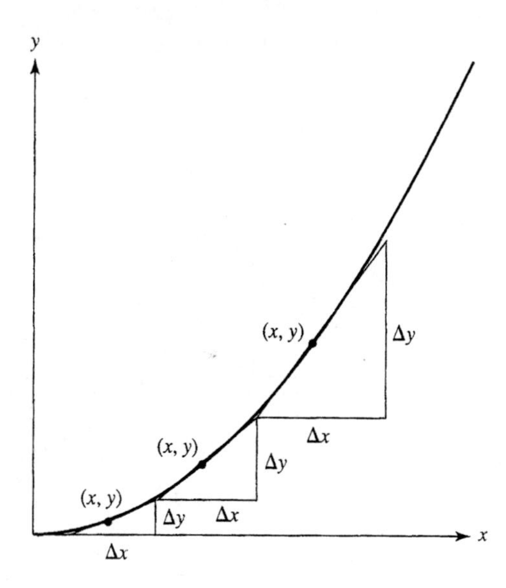

Figure 2-5 Variation of slope as a function of x.

identical roots in these cases will represent the zeros of the function. If the roots are imaginary, however, the equation will never even reach the x-axis, but will have a minimum (or a maximum, if the equation opens downward) above (or below) the x-axis. Under these circumstances, the roots will *not* represent the zeros of the function—the equation will have no real zeros.

To find the roots of a quadratic equation, we can use the *quadratic formula*

$$x = \frac{-b \pm \sqrt{b^2 - 4ac}}{2a} \tag{2-7}$$

Example

Find the roots of the equation $y = x^2 - x + 3$, using the quadratic formula. Are the roots the zeros of the function?

Solution

$$x = \frac{-(-1) \pm \sqrt{1 - 4(1)(3)}}{2} = \frac{1 \pm i\sqrt{11}}{2}$$

The roots are imaginary. The equation has no real zeros.

Sometimes, the roots of a quadratic equation can be found by factoring. For example, consider the equation

$$x^2 - 3x + 2 = 0 \tag{2-8}$$

which can be factored into the terms

$$(x - 1)(x - 2) = 0 \tag{2-9}$$

The roots of the equation now can be found by solving the equations

$$(x - 1) = 0 \quad \text{and} \quad (x - 2) = 0 \tag{2-10}$$

The result is $x = 1$ and $x = 2$. Substituting $a = 1, b = -3$, and $c = 2$ into the quadratic formula gives the same results.

In cases where the equation defining a particular physical situation is a second-degree equation (or even one of higher order), a problem arises that is not present when one simply considers the pure mathematics, as we have done. Since quadratic equations necessarily have two roots, we must decide, in cases where the roots are not the same, which root correctly represents the physical situation, even though both are mathematically correct. For example, consider the equilibrium equation

$$A + B \rightleftharpoons C + D$$

Assume that initially the concentration of A is 1.00 molar, the concentration of B is 2.00 molar and C and D are not present in the reaction vessel. Suppose we wish to determine the equilibrium concentrations of A, B, C, and D, given that the equilibrium constant for the reaction is 5.0. If we assume that, at equilibrium, the equilibrium concentrations of C and D are both equal to x, then, at equilibrium, the equilibrium concentrations of A and B will be $(1.00 - x)$ and $(2.00 - x)$, respectively. Substituting these expressions into the equilibrium constant equation

$$K_C = \frac{(C)(D)}{(A)(B)}$$

we have

$$5.0 = \frac{(x)(x)}{(1.00 - x)(2.00 - x)}$$

We can rearrange this equation to give the quadratic equation

$$4x^2 - 15x + 10 = 0$$

Using the quadratic formula, we obtain the roots: $x = 2.88$ and 0.868.

We now must decide which value of x is physically correct. If we choose $x = 2.88$, then the equilibrium concentrations of both A and B would be negative, which is physically impossible. Thus, the physically correct value for x must be 0.868, giving $0.132M$, $1.132M$, $0.868M$, and $0.868M$ for the equilibrium concentrations of A, B, C, and D, respectively. We see, then, that although both roots were mathematically correct, only one root is physically possible.

Exponential and Logarithmic Functions Exponential functions are functions whose defining equation is written in the general form

$$f(x) = a^x \tag{2-11}$$

where $a > 0$. An important exponential function that is used extensively in physical chemistry, and indeed in the physical sciences as a whole, is the function

$$f(x) = e^x \tag{2-12}$$

where the constant

$$e = \lim_{x \to 0}(1 + x)^{1/x} = 1 + \frac{1}{1!} + \frac{1}{2!} + \frac{1}{3!} + \cdots$$

is a nonterminating, nonrepeating decimal having the value 2.7183 to five significant figures. The function e^x is illustrated in Fig. 2-6. Note that all exponentials have the point (0, 1) in common, since $a^0 = 1$ for any a. Note also that there are no zeros to the function, since the function approaches zero as x approaches $-\infty$. The expression $\lim_{x \to 0}$, read "in the limit as x goes to zero," means that $(1 + x)^{1/x}$ approaches a value of 2.7183 as x approaches zero.

There is a direct relationship between exponential functions and logarithmic functions. The power or exponent to which the constant a is raised in the equation $y = a^x$ is called the *logarithm* of y to the base a and is written

$$\log_a y = x \tag{2-13}$$

The logarithmic function $\log_2 y = x$ is illustrated in Fig. 2-7. Note, as in the case of exponential functions, that the point (0, 1) is common to all logarithmic functions, since $\log_a 1 = 0$ for any a. Logarithms have many useful properties and are an important tool in the study of physical chemistry. For this reason, the general properties of logarithms are reviewed in Chapter 3.

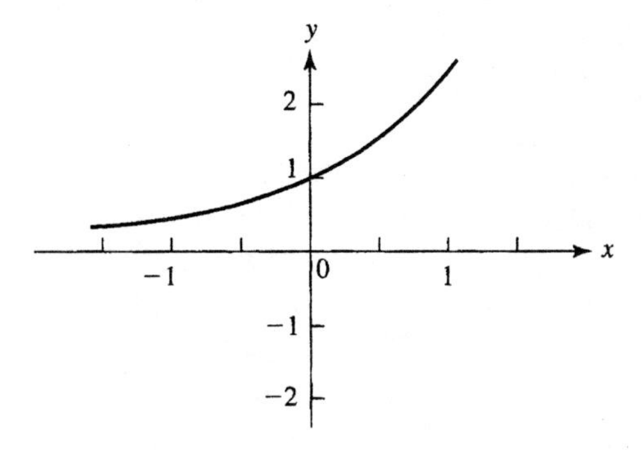

Figure 2-6 Graph of $y = e^x$.

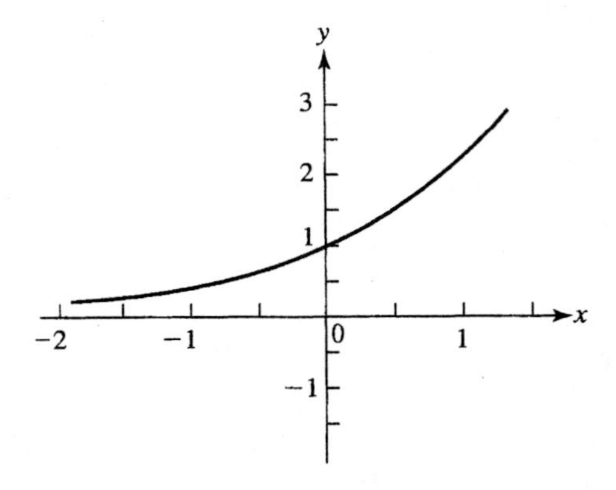

Figure 2-7 Graph of $\log_2 y = x$.

Circular Functions A circle is defined as the locus of all points in a plane that are at a constant distance from a fixed point. Circles are described by the equation

$$(x - a)^2 + (y - b)^2 = r^2 \tag{2-14}$$

where a and b are the coordinates of the center of the circle (the fixed point) and r is the radius. A unit circle is a circle with its center at the origin and a radius equal to unity:

$$x^2 + y^2 = 1 \tag{2-15}$$

Consider, now, the triangle inscribed in the unit circle, shown in Fig. 2-8. Let us define three functions: sine (abbreviated sin), which takes the angle θ into the y-coordinate of a point (x, y), cosine (abbreviated cos), which takes the angle θ into the x-coordinate of the point (x, y), and tangent (abbreviated tan), which is

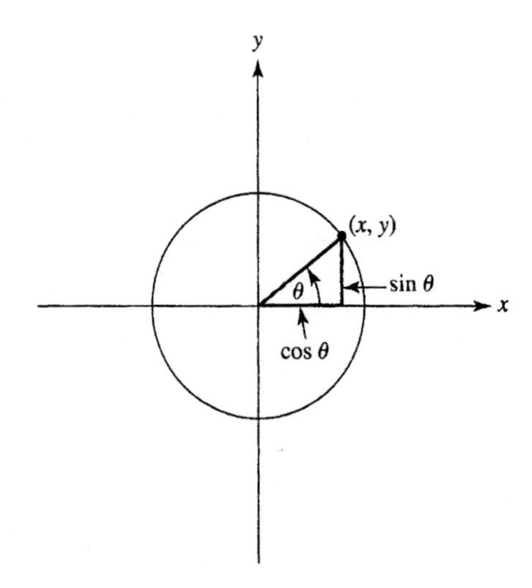

Figure 2-8 Graph of the unit circle $x^2 + y^2 = 1$.

the ratio of y to x. Thus,

$$\sin\theta = y \tag{2-16}$$

$$\cos\theta = x \tag{2-17}$$

$$\tan\theta = \frac{y}{x} = \frac{\sin\theta}{\cos\theta} \tag{2-18}$$

These functions are called *circular*, or *trigonometric*, functions. Note that Equations (2-16), (2-17), and (2-18) are just the transformation Equations (1-4) with $r = 1$. It is interesting to compare the graphs of functions, such as $\sin\theta$ and $\cos\theta$, in linear coordinates (coordinates in which θ is plotted linearly along the x-axis) with those in plane polar coordinates. Consider, for example, the equation $r = A\cos\theta$, where A is a constant. Such an equation can be used to describe the properties of waves and, indeed, describes p-type atomic orbitals in two dimensions. The functional dependence of r upon θ can be seen from the data listed in Table 2-1.

TABLE 2-1 DEPENDENCE OF r ON θ FOR THE FUNCTION $r = A\cos\theta$

θ (degrees)	r	θ (degrees)	r	θ (degrees)	r
0	1.000A	135	−0.707A	270	0
30	0.866A	150	−0.866A	300	0.500A
45	0.707A	180	−1.000A	315	0.707A
60	0.500A	210	−0.866A	330	0.866A
90	0	225	−0.707A	360	1.000A
120	−0.500A	240	−0.500A		

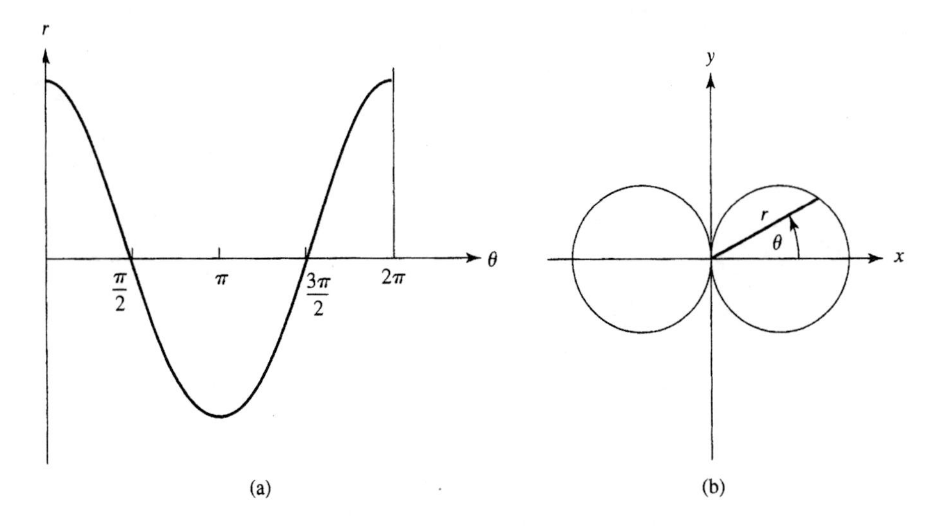

Figure 2-9 Graphs of $r = A\cos\theta$ plotted in (a) linear coordinates and (b) polar coordinates.

On the one hand, when r versus θ is plotted in linear coordinates [shown in Fig. 2-9(a)], the typical cosine curve results, certainly resembling a wave. On the other hand, when r versus θ is plotted in plane polar coordinates [shown in Fig. 2-9(b)], the more familiar shape of a p-orbital can be seen.[1] It is important to note that both graphs are equivalent; the shapes of the curves depend on the choice of coordinate system.

2-3 ROOTS OF POLYNOMIAL EQUATIONS

Earlier, we saw that the roots of a first- or second-degree equation can be found easily. But how do we find the roots of equations that are not linear or quadratic? Before the age of computers, this was not a simple task. One standard way to find the roots of a polynomial equation without using a computer is to graph the function. For example, consider the equation

$$y = x^4 + x^3 - 3x^2 - x + 1$$

If we plot this function from $x = -3$ to $x = +3$, we obtain the graph shown in Fig. 2-10. The roots of the equation are the values of x for which $y = 0$, or the

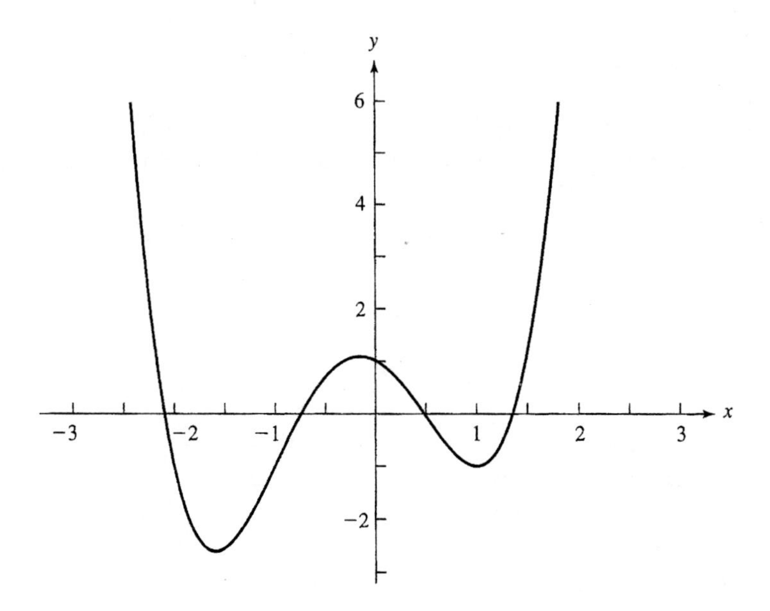

Figure 2-10 Graph of $y = x^4 + x^3 - 3x^2 - x + 1$.

[1]In polar coordinates, negative values of r have no physical meaning, so we are actually plotting $|r| = A\cos\theta$.

points on the graph where the graph crosses the x-axis. A careful examination of the graph shows that the roots are $x = -2.09$, $x = -0.74$, $x = +0.47$, and $x = +1.36$. In Chapter 11, we shall discuss numerical methods for finding roots of polynomial equations with the spreadsheet Excel.

SUGGESTED READINGS

1. BRADLEY, GERALD L., and SMITH, KARL J. *Calculus*. Upper Saddle River, NJ: Prentice Hall, 1995.
2. SULLIVAN, MICHAEL. *College Algebra*, 4th ed. Upper Saddle River, NJ: Prentice Hall, 1996.
3. VARBERG, DALE, and PURCELL, EDWIN J. *Calculus*, 7th ed. Upper Saddle River, NJ: Prentice Hall, 1997.
4. WASHINGTON, ALLYN J. *Basic Technical Mathematics*, 6th ed. Boston: Addison-Wesley, 1995.

PROBLEMS

1. Determine the roots of the following equations, and then state whether in each case the roots are the zeros of the function:

(a) $y = 5x - 10$
(b) $2(y - 2) = -6x$
(c) $y = x^2 - 2x + 4$
(d) $y = x^2 + x - 12$
(e) $y = x^2 - 1.509x - 2.922$
(f) $y = \cos x$
(g) $r = \sin\theta$
(h) $\mathrm{pH} = -\log_{10}(\mathrm{H}^+)$
(i) $x^2 + y^2 = 3$
(j) $(x - 3)^2 + (y - 4)^2 = 9$

2. Plot the following functions in plane polar coordinates from 0 to 2π (remember that, in polar coordinates, negative values of r have no physical meaning):

(a) $r = 2$
(b) $r = 4\sin\theta$
(c) $r = \theta/72$
(d) $r = 4\cos\theta$
(e) $r = 4\sin\theta\cos\theta$
(f) $r = 3\cos^2\theta - 1$

3. Plot the following functions in Cartesian coordinates:

(a) $y = 5$
(b) $y = 3x - 9$
(c) $y = 3e^{-x}$
(d) $y = 2\sin(2x)(0 \text{ to } 2\pi)$
(e) $PV = k$ (k is a constant)
(f) $\psi = 4\sin\theta\cos\theta$
(g) $y = x^2 - x + 3$
(h) $y = (4 - x^2)^{1/2}$

4. Plot the following functions, using the specified coordinates:

(a) $E_k = \frac{1}{2}mv^2$, (E_k versus v; m is constant)
(b) $V = -e^2/r$, (V versus r; e is a constant, not the exponential)
(c) $F = e^2/r^2$, (F versus r; e is a constant, not the exponential)
(d) $(A) = (A)_0 e^{-kt}$, ((A) versus t, $(A)_0$; k are constants; e is the exponential)
(e) $1/(A) = kt + 1/(A)_0$, ((A) versus t; k and $(A)_0$ are constants)
(f) $k_r = Ae^{-E/RT}$, (k_r versus T; A, E, and R are constants; e is the exponential)

5. Plot the functions in Problem 4, choosing suitable coordinates so that a straight line results.

6. Evaluate $[f(x + h) - f(x)]/h$ for the following functions:

(a) $f(x) = 1/x$ **(c)** $f(x) = 3x^2 - 3$

(b) $f(x) = 1/x^2$ **(d)** $f(x) = 1/(1 + x)$

7. Radioactive decay is a first-order process in which the concentration of the radioactive material C is related to time t by the equation

$$C = C_0 e^{-kt}$$

where C_0 and k are constants and e is the exponential function. Given the following data, determine the values of k and C_0 by plotting the data in such a way that a straight line results:

t(min)	0	30	60	90	120	150	180	210
C (counts/min)	3142	2601	2153	1782	1475	1221	1011	837

CHAPTER 3

Logarithms

3-1 INTRODUCTION

In the previous chapter, we defined a logarithm as the exponent or power x to which a number $a > 0$ is raised to give the number y. That is, if $a^x = y$, then $\log_a y$ (read "log of y to the base a") $= x$. Since the equations

$$a^x = y \quad \text{and} \quad \log_a y = x$$

are equivalent, we can use them to derive several useful properties of logarithms.

3-2 GENERAL PROPERTIES OF LOGARITHMS

PRODUCT RULE. *The logarithm of the product of two numbers m and n is equal to the sum of the logarithms of m and n:*

$$\log_a mn = \log_a m + \log_a n \tag{3-1}$$

Proof. Let $m = a^x$ and $n = a^y$. Then $x = \log_a m$ and $y = \log_a n$. Now,

$$mn = a^x a^y = a^{(x+y)} \tag{3-2}$$

Taking the logarithm to the base a of Equation (3-2) yields

$$\log_a mn = \log_a a^{(x+y)} = x + y = \log_a m + \log_a n$$

QUOTIENT RULE. *The logarithm of the quotient of two numbers m and n is equal to the difference of the logarithms of m and n:*

$$\log_a\left(\frac{m}{n}\right) = \log_a m - \log_a n \tag{3-3}$$

Proof. Let $m = a^x$ and $n = a^y$. Then $x = \log_a m$ and $y = \log_a n$. Now,

$$\frac{m}{n} = \frac{a^x}{a^y} = a^{(x-y)} \tag{3-4}$$

Taking the logarithm to the base a of Equation (3-4) produces

$$\log_a\left(\frac{m}{n}\right) = \log_a a^{(x-y)} = x - y = \log_a m - \log_a n$$

POWER RULE. *The logarithm of m raised to the power n is equal to n multiplied by the logarithm of m.*

$$\log_a(m)^n = n \log_a m \tag{3-5}$$

Proof. Let $m = a^x$. Then $x = \log_a m$. Now,

$$m^n = (a^x)^n = a^{xn} \tag{3-6}$$

Taking the logarithm to the base a of Equation (3-6) gives

$$\log_a(m)^n = \log_a a^{nx} = nx = n \log_a m$$

3-3 COMMON LOGARITHMS

In the previous examples, we did not specify any particular value for the base a; that is, the preceding rules hold for any value of a. In numerical calculations, however, we find it convenient to use logarithms to the base 10, since they are directly related to our decimal system of expressing numbers and also are linked to what we normally refer to as *scientific notation,* in which we express numbers in terms of powers of 10. Such logarithms are called *common logarithms* and are written simply as $\log y$. The relationship between exponents of the number 10 and common logarithms can be seen in Table 3-1.

TABLE 3-1 RELATIONSHIP BETWEEN $y = 10^x$ AND $\log_{10} y = x$

$10^0 = 1$	$\log 1 = 0$	$10^{0.3010} = 2$	$\log 2 = 0.3010$
$10^1 = 10$	$\log 10 = 1$	$10^{0.4771} = 3$	$\log 3 = 0.4771$
$10^2 = 100$	$\log 100 = 2$	$10^{0.9542} = 9$	$\log 9 = 0.9542$
$10^3 = 1000$	$\log 1000 = 3$	$10^{1.3010} = 20$	$\log 20 = 1.3010$

In general, a logarithm is composed of two parts: a *mantissa*, which is a positive number that determines the exact value of the number from 1 to 9.999 ..., and a *characteristic* (multiplier), which can be positive or negative and which determines where the decimal point is placed in the number. Using logarithms is equivalent to expressing all numbers in scientific notation—for example, 12,200 as 1.22×10^4. The number 1.22 is equivalent to the mantissa of the logarithm, and the 10^4, which tells us where the decimal point is placed, is equivalent to the characteristic of the logarithm. In fact, if we determine the logarithm of 12,200, we see that it is equal to the logarithm of 1.22 plus the logarithm of 10^4:

$$\log(12200) = \log(1.22) + \log(10^4) = 0.0864 + 4 = 4.0864$$

Here, 0.0864 is the mantissa and 4 is the characteristic. It is important to note that the number of significant figures in a number is related to the mantissa and not the characteristic. The number 12,200 has three significant figures, and the mantissa reflects that fact. The characteristic 4 in the logarithm tells us only where the decimal point is placed.

A negative characteristic designates a number lying in a range $0 < N < 1$. To emphasize the fact that a particular logarithm is made up of a negative characteristic and a positive mantissa (mantissas are never negative), a minus sign is placed above the characteristic. Thus, $\log 0.020 = \log(2.0 \times 10^{-2})$ is expressed as $\bar{2}.3010$. Such a logarithm is called a ***heterogeneous logarithm***. It is possible to combine the negative characteristic with the positive mantissa to form a *homogeneous logarithm*. Calculators and computers do so automatically. When this is done, the negative sign is placed in front of the logarithm, so that $\bar{2}.3010 = -1.6990$. The importance of the homogeneous logarithm to physical chemistry lies in the fact that all logarithmic data and certain physical quantities, such as the pH of a substance, are expressed in homogeneous form, since logarithms expressed in heterogeneous form could never conveniently be scaled on a graphical axis.

Before the age of hand calculators, common logarithms were used extensively to perform many types of calculations that today seem rather trivial, such as determining the roots of a number. (Try to find the fifth root of 43.9987 without the use of a calculator!) In fact, determining the logarithm of a number itself required wading through tables of numbers. But the calculator has changed all that. Today, we can easily determine the logarithm of a number simply by pressing a key on our calculator. We find, however, that while today it may not be necessary to use logarithms to multiply numbers together or to find roots of numbers, logarithms are still important, since a number of important chemical concepts, such as pH and optical absorbance, are defined in terms of the common, or base-10, logarithms. So it is important for students of physical chemistry to become familiar with the log key (and the inverse, or antilog, key) on their calculators.

3-4 NATURAL LOGARITHMS

In Chapter 2, we introduced the function $f(x) = e^x$ as an exponential function that is particularly important in the study of physical chemistry. Logarithms taken to the base e are known as *natural logarithms* and are designated $\ln y = x$. Before going into the physical significance of the natural logarithm, it might be useful to consider the relationship between natural and common logarithms. Consider the equation

$$y = e^x \tag{3-7}$$

Taking the logarithm to the base 10 of this equation gives

$$\log y = \log(e^x) = x \log e \tag{3-8}$$

But $x = \ln y$, so, substituting this into Equation (3-8), we get

$$\log y = \log e \ln y = \log(2.718) \ln y = 0.4343 \ln y$$

or

$$\ln y = 2.303 \log y \tag{3-9}$$

The physical significance of the natural logarithm can best be explained by considering the fractional change in any variable y, written $\Delta y/y$, where Δy is some finite change in y. Suppose that a variable y changes to a new value $y + \Delta y$. If the change Δy is small, then the change in the natural logarithm of y is

$$\begin{aligned}\Delta \ln y &= \ln(y + \Delta y) - \ln y \\ &= \ln\left(\frac{y + \Delta y}{y}\right) = \ln\left(1 + \frac{\Delta y}{y}\right)\end{aligned}$$

Dividing both sides by Δy gives

$$\begin{aligned}\frac{\Delta \ln y}{\Delta y} &= \frac{1}{\Delta y} \ln\left(1 + \frac{\Delta y}{y}\right) \\ &= \ln\left(1 + \frac{\Delta y}{y}\right)^{1/\Delta y} = \frac{1}{y} \ln\left(1 + \frac{\Delta y}{y}\right)^{y/\Delta y}\end{aligned}$$

Remember from Chapter 2, however, that the exponential e is defined as $\lim_{x \to 0}(1 + x)^{1/x}$. If we let $x = \Delta y/y$, then it follows that

$$\lim_{\Delta y \to 0}\left(1 + \frac{\Delta y}{y}\right)^{y/\Delta y} = \lim_{x \to 0}(1 + x)^{1/x} = e$$

$$\frac{\Delta \ln y}{\Delta y} = \frac{1}{y} \ln e = \frac{1}{y} \tag{3-10}$$

since $\ln e = 1$. Rearranging Equation (3-10) gives

$$\lim_{\Delta y \to 0} \frac{\Delta y}{y} = \Delta \ln y \tag{3-11}$$

In general, then, we see that, in the limit such that the change in any variable is vanishingly small, the *fractional change* in that variable is equal to the change in the natural logarithm of the variable. The mathematical significance of Equation (3-11) will be explored in Chapter 4.

SUGGESTED READINGS

1. BRADLEY, GERALD L., and SMITH, KARL J. *Calculus*. Upper Saddle River, NJ: Prentice Hall, 1995.
2. SULLIVAN, MICHAEL. *College Algebra*, 4th ed. Upper Saddle River, NJ: Prentice Hall, 1996.
3. VARBERG, DALE, and PURCELL, EDWIN J., *Calculus*, 7th ed. Upper Saddle River, NJ: Prentice Hall, 1997.

PROBLEMS

1. The apparent pH of an aqueous solution is defined as $\text{pH} = -\log_{10}(\text{H}^+)$. Find the apparent pH of the following solutions:

(a) $(\text{H}^+) = 1.00 \times 10^{-7}\ M$
(b) $(\text{H}^+) = 0.123\ M$
(c) $(\text{H}^+) = 8.54 \times 10^{-10}\ M$
(d) $(\text{H}^+) = 1.152\ M$
(e) $(\text{H}^+) = 6.23 \times 10^{-2}\ M$
(f) $(\text{H}^+) = 12.0\ M$

2. Given the following values for the apparent pH, find (H^+) for the following solutions:

(a) pH = 0
(b) pH = 2.156
(c) pH = 0.234
(d) pH = 7.876
(e) pH = 13.435
(f) pH = −0.132

3. Find the pH of an aqueous solution of HCl in which the concentration of HCl is $2.00 \times 10^{-8}\ M$. (*Hint*: The pH cannot be above 7.00, because the solution is acidic. Do not forget the dissociation of water!)

4. The work done in an isothermal, reversible expansion or compression of an ideal gas from a volume V_1 to a volume V_2 is given by the equation

$$w = -nRT \ln \frac{V_2}{V_1}$$

where n is the number of moles of the gas, R is the gas constant $= 8.314\ \text{J} \cdot \text{mol}^{-1} \cdot \text{K}^{-1}$, and T is the absolute temperature. Find the work done in the isothermal, reversible expansion of 1.00 mole of an ideal gas at 298.2K from a volume of 3.00 liters to a volume of 10.00 liters.

5. The change in entropy associated with the expansion or compression of an ideal gas is given by the equation

$$\Delta S = nC_V \ln\frac{T_2}{T_1} + nR \ln\frac{V_2}{V_1}$$

where n is the number of moles of the gas, C_V is the molar heat capacity of the gas at constant volume, T is the absolute temperature of the gas, and V is the volume of the gas. Find the change in entropy attending the expansion of 1.00 mole of an ideal gas from 1.00 liter to 5.00 liters if the temperature drops from 300 K to 284 K. Take $C_V = \frac{3}{2}R$ and $R = 8.314\ \mathrm{J \cdot mol^{-1} \cdot K^{-1}}$.

6. If the expansion of an ideal gas takes place adiabatically and reversibly, then ΔS of Problem 5 is zero. Using the equation given in that problem, find the final temperature of an ideal gas ($C_V = 3/2\ R$) that is initially at 300 K and that expands adiabatically and reversibly from 1.00 liter to 5.00 liters.

7. Radioactive decay is a first-order process that follows the integrated rate law

$$\ln\frac{(A)}{(A)_0} = -kt$$

where (A) is the concentration of A at time t, $(A)_0$ is the initial concentration of A (i.e., the concentration at $t = 0$), and k is a constant called the *rate constant*. The fraction of ^{14}C in a sample of wood ash from an archeological dig was found to be 0.664. How old is the wood ash, given that $k = 1.24 \times 10^{-4}\ \mathrm{yr^{-1}}$ for the radioisotope ^{14}C.

8. The EMF of a voltaic cell, E_{cell}, is related to the temperature and concentrations of ions in the cell reaction $a\mathrm{A} + b\mathrm{B} \rightarrow c\mathrm{C} + d\mathrm{D}$ by the Nernst equation

$$E_{cell} = E^0_{cell} - \frac{RT}{nF} \ln\frac{(\mathrm{C})^c(\mathrm{D})^d}{(\mathrm{A})^a(\mathrm{B})^b}$$

where E^0_{cell} is the EMF of the cell when all concentrations are 1.00 M (known as the standard EMF), $R = 8.314\ \mathrm{J \cdot mol^{-1} \cdot K^{-1}}$, T is the absolute temperature, n is the number of electron change in the cell reaction, and F is the Faraday constant 96,487 coulombs/equivalent. Find the EMF of a cell at 298.2K for the reaction

$$\mathrm{Zn(s) + Cu^{++} \rightarrow Zn^{++} + Cu(s)}$$

when $(\mathrm{Cu^{++}}) = 0.500\ M$ and $(\mathrm{Zn^{++}}) = 0.100\ M$, given that $E^0_{cell} = 1.100$ V.

9. Verify quantitatively that as the change in x gets smaller and smaller, $\Delta x/x$ approaches $\Delta \ln x$.

CHAPTER 4

Differential Calculus

4-1 INTRODUCTION

Physical chemistry is concerned to a great extent with the effect that a change in one variable of a system will have on the other variables of the system. For example, how will a change in the pressure or temperature of a gas affect its volume or energy? *Differential calculus* is the mathematics of incremental changes. It is based primarily on the mathematical concept known as the *derivative*. The derivative of a variable y with respect to a variable x, where y must be a function of x, is defined as

$$\frac{dy}{dx} = \lim_{\Delta x \to 0} \frac{\Delta y}{\Delta x} \tag{4-1}$$

where Δy and Δx denote changes in the variables y and x, respectively. Thus, the derivative of y with respect to x is simply the change in $f(x)$ with respect to the change in x when the change in x becomes vanishingly small. If y is not a function of x, then the derivative does not exist (i.e., dy/dx is equal to zero). We saw in a previous chapter that the slope of a curve at a point can be defined by a line drawn tangent to the curve at that point. We shall see in a subsequent section that if (x_1, y_1) and (x_2, y_2) are two points on the curve, $\Delta y/\Delta x$ more closely approximates the tangent line at (x_1, y_1) as Δx gets smaller and smaller. Therefore, in the limit as Δx goes to zero, $\Delta y/\Delta x = dy/dx$ is the slope of the curve at the point (x_1, y_1).

It is important to emphasize that, while mathematically it might be a more or less straightforward procedure to take the derivative of a function once the equation describing the functional dependence is known, it is the job of the scientist to determine how one variable of a system depends on other variables and to find the equation relating them. Some scientists are exceptionally good at doing this and win Nobel prizes; the rest of us keep trying. That is why it is so important, not only to understand the mathematics, but also to learn and understand the science. For example, students insist on describing the isothermal (constant-temperature) expansion of a gas held by a piston–cylinder arrangement (see Fig. 4-3) against a constant external pressure as an isobaric (constant-pressure) process. If a gas, ideal or otherwise, expands at constant temperature, its pressure *has* to change. (See Boyle's law for an ideal gas!) There is no functional dependence between the external pressure on the gas (part of the surroundings) and the volume of the gas (part of the system). Thus, no derivative exists. Only in the very special case of the reversible expansion or compression of a gas, which we will discuss in a subsequent section, can they be related, and then only indirectly.

The derivative of a function may be taken more than once, giving rise to second, third, and higher derivatives, denoted d^2y/dx^2, d^3y/dx^3, and so on. Note that the second derivative is the first derivative of the first derivative, that is,

$$\frac{d}{dx}\left(\frac{dy}{dx}\right) = \frac{d^2y}{dx^2}$$

and the third derivative is the first derivative of the second derivative, or

$$\frac{d}{dx}\left(\frac{d^2y}{dx^2}\right) = \frac{d^3y}{dx^3}$$

and so on. The process of taking derivatives is called *differentiation.*

Differential calculus has many uses in physical chemistry; however, before going into these, let us first review the mechanics of differentiation. The functional dependence of the variables of a system may appear in many different forms: as first- or second-degree equations, as trigonometric functions, as logarithms, or as exponential functions. For this reason, we need to know the derivatives of these types of functions, which are used extensively in physical chemistry. Also included in the list of functions and their derivatives in the next section are rules for differentiating sums, products, and quotients. In some cases, examples are given in order to illustrate the application to physicochemical equations.

4-2 FUNCTIONS OF SINGLE VARIABLES

1. $\frac{d}{dx}(c) = 0,$ where c is any quantity that is not a function of x (e.g., a constant).

2. $\frac{d}{dx}(cx) = c,$ where c is any quantity that is not a function of x.

Example: $V = kT; \frac{dV}{dT} = k = \frac{V}{T}$

3. $\frac{d}{dx}(x^n) = nx^{n-1},$ where n is any real number.

Examples:

(a) $P = \frac{k}{V} = kV^{-1}; \frac{dP}{dV} = (-1)kV^{-2} = -\frac{k}{V^2} = -\frac{P}{V}$

$$\frac{d^2P}{dV^2} = (-2)(-1)kV^{-3} = \frac{2k}{V^3}$$

(b) The derivative of the volume of a sphere with respect to its radius is the surface area of the sphere:

$$V = \tfrac{4}{3}\pi r^3; \quad \frac{dV}{dr} = (3)\left(\frac{4}{3}\right)\pi r^2 = 4\pi r^2$$

(c) The derivative of kinetic energy with respect to velocity is momentum:

$$E_k = \tfrac{1}{2}mv^2; \quad \frac{dE_k}{dv} = 2\left(\frac{1}{2}\right)mv = mv$$

(d) $r = \sqrt{x} = (x)^{1/2}; \quad \dfrac{dr}{dx} = \dfrac{1}{2}x^{-1/2}$

4. The derivative of a sum or difference of terms is the sum or difference of the derivatives of the terms:

$$\frac{d}{dx}[f(x) + g(x)] = \frac{df}{dx} + \frac{dg}{dx} \quad \text{and} \quad \frac{d}{dx}[f(x) - g(x)] = \frac{df}{dx} - \frac{dg}{dx}$$

Examples:

(a) $y = mx + b$, where m and b are constants.

$$\frac{dy}{dx} = \frac{d(mx)}{dx} + \frac{db}{dx} = m + 0 = m$$

(b) $\ln P = \dfrac{-\Delta H}{RT} + c$, where ΔH, R, and c are constants.

$$\frac{d(\ln P)}{dT} = \frac{\Delta H}{RT^2}$$

5. $\dfrac{d}{dx}(\sin ax) = a\cos ax$, where a is a constant.

Examples:

(a) $\psi = A\sin\dfrac{n\pi x}{a}$, where A, n, π, and a are constants.

$$\frac{d\psi}{dx} = \left(\frac{n\pi}{a}\right)A\cos\left(\frac{n\pi x}{a}\right)$$

(b) $y = A\sin(2\pi\nu t)$, where A, π, and ν are constants.

$$\frac{dy}{dt} = (2\pi\nu)A\cos(2\pi\nu t)$$

6. $\dfrac{d}{dx}(\cos ax) = -a \sin ax,$ where a is a constant.

Examples:

(a) $y = A\cos(2\pi\nu t); \dfrac{dy}{dt} = -2\pi\nu A \sin(2\pi\nu t)$

(b) $$\frac{d^2}{dx^2}\left[A \sin\left(\frac{n\pi x}{a}\right)\right] = \frac{d}{dx}\left[\left(\frac{n\pi}{a}\right)A\cos\left(\frac{n\pi x}{a}\right)\right]$$
$$= -\frac{n^2\pi^2}{a^2}A\sin\left(\frac{n\pi x}{a}\right)$$

7. $\dfrac{d}{dx}(\tan x) = \sec^2 x$

8. $\dfrac{d}{dx}(e^{ax}) = ae^{ax},$ where a is a constant.

Examples:

(a) $\Phi = Ae^{im\phi},$ where A and m are constants and $i = \sqrt{-1}$.

$$\frac{d\Phi}{d\phi} = imAe^{im\phi}$$

$\dfrac{d^2\Phi}{d\phi^2} = -m^2Ae^{im\phi} = -m^2\Phi$. This is the ϕ equation describing the stationary states of the hydrogen atom.

(b) $(A) = (A)_0e^{-kt},$ where $(A)_0$ and k are constants.

$$\frac{d(A)}{dt} = -k(A)_0e^{-kt}$$

9. $\dfrac{d}{dx}(\ln x) = \dfrac{1}{x}$ or $\dfrac{dx}{x} = d(\ln x)$. This equation is identical to Equation (3-11) and emphasizes the physical significance of the natural logarithm.

10. $\dfrac{d}{dx}[f(g(x))] = \dfrac{df}{dg}\cdot\dfrac{dg}{dx}$. This rule of differentiation is known as the *chain rule.* It is one of most used rules of differentiation and finds many applications in physical chemistry.

Examples:

(a) $\Phi = 3\cos^2\theta - 1.$

Let $u = \cos\theta; \dfrac{du}{d\theta} = -\sin\theta$

$$\Phi = 3u^2 - 1; \frac{d\Phi}{du} = 6u$$

$$\frac{d\Phi}{d\theta} = \frac{d\Phi}{du}\cdot\frac{du}{d\theta} = -6u \sin\theta = -6\cos\theta \sin\theta$$

(b) $y = \dfrac{1}{\sqrt{1-x^2}} = (1-x^2)^{-1/2}$

Let $u = (1-x^2); \dfrac{du}{dx} = -2x$, and $y = u^{-1/2}$.

$$\frac{dy}{du} = -\frac{1}{2}u^{-3/2}; \frac{dy}{dx} = \frac{dy}{du}\cdot\frac{du}{dx}$$

$$\frac{dy}{dx} = -\frac{1}{2}(1-x^2)^{-3/2}(-2x) = \frac{x}{(1-x^2)^{3/2}}$$

(c) $n = n_0 e^{-E/RT}$, where n_0, E, and R are constants.

Let $u = E/RT; \dfrac{du}{dT} = -\dfrac{E}{RT^2}$

$$n = n_0 e^{-u}; \frac{dn}{du} = -n_0 e^{-u}$$

$$\frac{dn}{dT} = \frac{dn}{du}\cdot\frac{du}{dT} = -n_0 e^{-E/RT}\left(-\frac{E}{RT^2}\right) = n_0\left(\frac{E}{RT^2}\right)e^{-E/RT}$$

(d) $y = Ae^{-ax^2}$, where A and a are constants.

Let $u = ax^2; \dfrac{du}{dx} = 2ax$

$$y = Ae^{-u}; \frac{dy}{du} = -Ae^{-u}$$

$$\frac{dy}{dx} = \frac{dy}{du}\cdot\frac{du}{dx} = -Ae^{-ax^2}(2ax) = -2axAe^{-ax^2}$$

11. $\dfrac{d}{dx}[f(x)\cdot g(x)] = f(x)\dfrac{dg}{dx} + g(x)\dfrac{df}{dx}$

Examples:

(a) $y = (\sin x)e^{mx}$, where m is a constant.
Let $f(x) = \sin x$ and $g(x) = e^{mx}$

$$\frac{df}{dx} = \cos x \quad \text{and} \quad \frac{dg}{dx} = me^{mx}$$

$$\frac{dy}{dx} = m(\sin x)e^{mx} + (\cos x)e^{mx}$$

(b) $F = -\eta 2\pi r L \dfrac{dy}{dr}$, where η, π, and L are constants.

Let $f(r) = -\eta 2\pi r L$ and $g(r) = \dfrac{dy}{dr}$

$$\frac{df}{dr} = -\eta 2\pi L \quad \text{and} \quad \frac{dg}{dr} = \frac{d^2y}{dr^2}$$

$$\frac{dF}{dr} = -\eta 2\pi r L \frac{d^2y}{dr^2} - \eta 2\pi L \frac{dy}{dr}$$

(c) $E = kT^2 \dfrac{d}{dT}(\ln q)$, where k is a constant.

Let $f(T) = kT^2$ and $g(T) = \dfrac{d}{dT}(\ln q)$

$$\frac{df}{dT} = 2kT \quad \text{and} \quad \frac{dg}{dT} = \frac{d^2}{dT^2}(\ln q)$$

$$\frac{dE}{dT} = kT^2 \frac{d^2}{dT^2}(\ln q) + 2kT \frac{d}{dT}(\ln q)$$

(d) $\psi(x) = e^{-x^2/2} y(x)$

Let $u = -\dfrac{x^2}{2}$; $\dfrac{du}{dx} = -x$; $\dfrac{d^2u}{dx^2} = -1$. Therefore, $\psi(x) = e^u y(x)$

$$\frac{d\psi}{dx} = e^u \frac{dy}{dx} + y(x)\frac{d(e^u)}{dx} = e^u \frac{dy}{dx} + y(x)e^u \frac{du}{dx}$$

$$\frac{d^2\psi}{dx^2} = e^u \frac{d^2y}{dx^2} + \frac{dy}{dx} e^u \frac{du}{dx} + y(x)\left[e^u \frac{d^2u}{dx^2} + \frac{du}{dx} e^u \frac{du}{dx}\right] + e^u \frac{du}{dx}\frac{dy}{dx}$$

$$\frac{d^2\psi}{dx^2} = e^{-x^2/2} \frac{d^2y}{dx^2} - xe^{-x^2/2} \frac{dy}{dx} - y(x)e^{-x^2/2}$$

$$+ x^2 y(x) e^{-x^2/2} - xe^{-x^2/2} \frac{dy}{dx}$$

$$= e^{-x^2/2} \frac{d^2y}{dx^2} - 2xe^{-x^2/2} \frac{dy}{dx} + x^2 y(x) e^{-x^2/2} - y(x)e^{-x^2/2}$$

12. $$\frac{d}{dx}\left[\frac{f(x)}{g(x)}\right] = \frac{g(x)\dfrac{df}{dx} - f(x)\dfrac{dg}{dx}}{(g(x))^2}$$

Example: $y = \tan x = \dfrac{\sin x}{\cos x}$

Let $f(x) = \sin x$ and $g(x) = \cos x$

$$\frac{df}{dx} = \cos x \quad \text{and} \quad \frac{dg}{dx} = -\sin x$$

$$\frac{dy}{dx} = \frac{\cos x(\cos x) - \sin x(-\sin x)}{\cos^2 x} = \frac{\cos^2 x + \sin^2 x}{\cos^2 x}$$

$$= \frac{1}{\cos^2 x} = \sec^2 x$$

4-3 FUNCTIONS OF SEVERAL VARIABLES; PARTIAL DERIVATIVES

In the previous section, the functions that were differentiated were functions of only one independent variable. Most physicochemical systems, however, normally encompass more than one independent variable. For example, the pressure of an ideal gas is simultaneously a function of the temperature and the volume of the gas. This relationship can be expressed in the form of an equation of state for the gas, namely,

$$P = f(T, V) = \frac{RT}{V} \tag{4-2}$$

where R is a constant. Since both independent variables can change, let us consider two ways to treat this situation. First we consider the case where only one of the variables changes while the other remains constant. The derivative of P with respect to only one of the variables T or V while the other remains constant is called a *partial derivative* and is designated by the symbol ∂. The partial derivative of P with respect to V at constant T can be defined as

$$\left(\frac{\partial P}{\partial V}\right)_T = \lim_{\Delta V \to 0} \frac{f(T, V + \Delta V) - f(T, V)}{\Delta V} \tag{4-3}$$

and the partial derivative of P with respect to T at constant V can be defined as

$$\left(\frac{\partial P}{\partial T}\right)_V = \lim_{\Delta T \to 0} \frac{f(T + \Delta T, V) - f(T, V)}{\Delta T} \tag{4-4}$$

The small subscripts T and V attached to the expressions $\partial P/\partial V$ and $\partial P/\partial T$, respectively, indicate which variables are to be held constant.

The rules for partial differentiation are the same as those for ordinary differentiation (see Section 4-2), with the addition that the variables held constant are treated the same as the other constants in the equation. Hence, at constant T,

$$P = \frac{RT}{V}; \left(\frac{\partial P}{\partial V}\right)_T = \frac{-RT}{V^2} \tag{4-5}$$

and at constant V,

$$P = \frac{RT}{V}; \left(\frac{\partial P}{\partial T}\right)_V = \frac{R}{V} \tag{4-6}$$

Functions of two or more variables can be differentiated partially more than once with respect to either variable while holding the other constant, to yield second and higher order derivatives. For example,

$$\left(\frac{\partial}{\partial T}\left(\frac{\partial P}{\partial T}\right)_V\right)_V = \left(\frac{\partial^2 P}{\partial T^2}\right)_V \tag{4-7}$$

and

$$\left(\frac{\partial}{\partial V}\left(\frac{\partial P}{\partial T}\right)_V\right)_T = \left(\frac{\partial^2 P}{\partial V \partial T}\right) \tag{4-8}$$

Equation (4-8) is called a *mixed partial second derivative.* If a function of two or more variables and its derivatives are single valued and continuous—two properties normally attributed to physical variables—then the function's mixed partial second derivatives are equal. That is,

$$\left(\frac{\partial}{\partial V}\left(\frac{\partial P}{\partial T}\right)_V\right)_T = \left(\frac{\partial}{\partial T}\left(\frac{\partial P}{\partial V}\right)_T\right)_V \tag{4-9}$$

or

$$\left(\frac{\partial^2 P}{\partial V \partial T}\right) = \left(\frac{\partial^2 P}{\partial T \partial V}\right) \tag{4-10}$$

To illustrate that partial differentiation is, in fact, no more complicated than ordinary differentiation, consider the following examples:

Examples

(a) $d = \frac{m}{V}; \left(\frac{\partial d}{\partial m}\right)_V = \frac{1}{V}; \left(\frac{\partial d}{\partial V}\right)_m = \frac{-m}{V^2}$

(b) $V = \pi r^2 h; \left(\frac{\partial V}{\partial r}\right)_h = 2\pi r h; \left(\frac{\partial V}{\partial h}\right)_r = \pi r^2$

(c) $\left(\frac{\partial E}{\partial T}\right)_V = T\left(\frac{\partial S}{\partial T}\right)_V$

Take the second derivative of E with respect to V at constant T:

$$\left(\frac{\partial}{\partial V}\left(\frac{\partial E}{\partial T}\right)_V\right)_T = T\left(\frac{\partial}{\partial V}\left(\frac{\partial S}{\partial T}\right)_V\right)_T = T\left(\frac{\partial^2 S}{\partial V \partial T}\right)$$

(d) $\left(\frac{\partial E}{\partial V}\right)_T = T\left(\frac{\partial S}{\partial V}\right)_V - P,$ where $P = f(T, V)$

Take the second derivative of E with respect to T at constant V.

Solution. Because both T and $(\partial S/\partial V)$ are functions of V, the first term must be differentiated as a product:

$$\left(\frac{\partial}{\partial T}\left(\frac{\partial E}{\partial V}\right)_T\right)_V = T\left(\frac{\partial}{\partial T}\left(\frac{\partial S}{\partial V}\right)_T\right)_V + \left(\frac{\partial S}{\partial V}\right)_T\left(\frac{\partial T}{\partial T}\right) - \left(\frac{\partial P}{\partial T}\right)_V$$

Since $(\partial T/\partial T) = 1$, we can write

$$\left(\frac{\partial^2 E}{\partial T \partial V}\right) = T\left(\frac{\partial^2 S}{\partial T \partial V}\right) + \left(\frac{\partial S}{\partial V}\right)_T - \left(\frac{\partial P}{\partial T}\right)_V$$

(e) $\psi(r, \theta, \phi) = R(r)\Theta(\theta)\Phi(\phi) = R(r)\Theta(\theta)Ae^{\pm im\phi}$

$$\frac{\partial \psi}{\partial \phi} = R(r)\Theta(\theta)(\pm im)Ae^{\pm im\phi}$$

$$\frac{\partial^2 \psi}{\partial \phi^2} = R(r)\Theta(\theta)(-m^2)Ae^{\pm im\phi} = -m^2 R(r)\Theta(\theta)\Phi(\phi)$$

$$= -m^2\psi(r, \theta, \phi)$$

(f) Convert the partial-derivative operator $\partial/\partial x$ to plane polar coordinates, using the *chain rule*.

Solution. Using the chain rule, we get

$$\frac{\partial}{\partial x} = \frac{\partial r}{\partial x}\frac{\partial}{\partial r} + \frac{\partial \theta}{\partial x}\frac{\partial}{\partial \theta}$$

The reverse-transformation equations for plane polar coordinates are (see Chapter 1)

$$r = (x^2 + y^2)^{1/2} \quad \text{and} \quad \theta = \tan^{-1}\left(\frac{y}{x}\right)$$

so

$$\frac{\partial r}{\partial x} = \frac{1}{2}(x^2 + y^2)^{-1/2}(2x) = \frac{x}{r} = \frac{r\cos\theta}{r} = \cos\theta$$

To find $\partial\theta/\partial x$, let $u = (y/x)$. Then $\dfrac{\partial u}{\partial x} = \dfrac{-y}{x^2}$,

$$\theta = \tan^{-1}u; \quad \frac{\partial\theta}{\partial u} = \frac{1}{1+u^2} = \frac{1}{1+(y/x)^2} = \frac{x^2}{x^2+y^2}$$

and

$$\frac{\partial\theta}{\partial x} = \frac{\partial\theta}{\partial u}\cdot\frac{\partial u}{\partial x} = \left(\frac{x^2}{x^2+y^2}\right)\left(\frac{-y}{x^2}\right) = \frac{-r\sin\theta}{r^2} = \frac{-\sin\theta}{r}$$

Therefore,

$$\frac{\partial}{\partial x} = \cos\theta\frac{\partial}{\partial r} - \frac{\sin\theta}{r}\frac{\partial}{\partial\theta}$$

4-4 THE TOTAL DIFFERENTIAL

We now consider the second case in which the independent variables of a system may be varied; we examine the effect of this variation on the dependent variable. In the previous section, we allowed only one variable to change at a time. In this section, we shall consider the effect of allowing all of the variables to change simultaneously. Consider again the example $P = f(T, V)$. Let ΔP represent the change in pressure brought about by a simultaneous change in temperature and volume. That is,

$$\Delta P = f(T + \Delta T, V + \Delta V) - f(T, V) \tag{4-11}$$

Adding and subtracting $f(T, V + \Delta V)$ to Equation (4-11) yields

$$\begin{aligned}\Delta P = {} & f(T + \Delta T, V + \Delta V) - f(T, V + \Delta V) \\ & + f(T, V + \Delta V) - f(T, V)\end{aligned} \tag{4-12}$$

Multiplying the first two terms in Equation (4-12) by $\Delta T/\Delta T$ and the second two terms by $\Delta V/\Delta V$ gives

$$\begin{aligned}\Delta P = {} & \left[\frac{f(T + \Delta T, V + \Delta V) - f(T, V + \Delta V)}{\Delta T}\right]\Delta T \\ & + \left[\frac{f(T, V + \Delta V) - f(T, V)}{\Delta V}\right]\Delta V\end{aligned} \tag{4-13}$$

Taking the limit as ΔT goes to zero at constant V for the first term and the limit as ΔV goes to zero at constant T for the second term results in

$$\lim_{\Delta P \to 0} \Delta P = \lim_{\Delta T \to 0}\left[\frac{f(T + \Delta T, V) - f(T, V)}{\Delta T}\right]\Delta T + \lim_{\Delta V \to 0}\left[\frac{f(T, V + \Delta V) - f(T, V)}{\Delta V}\right]\Delta V \tag{4-14}$$

The terms in brackets are just the partial derivatives $(\partial P/\partial T)_V$ and $(\partial P/\partial V)_T$. Replacing ΔP, ΔT, and ΔV with dP, dT, and dV, respectively, to indicate vanishingly small changes, we can write

$$dP = \left(\frac{\partial P}{\partial T}\right)_V dT + \left(\frac{\partial P}{\partial V}\right)_T dV \tag{4-15}$$

where the expression dP represents the *total differential* of P. The terms $(\partial P/\partial T)_V\, dT$ and $(\partial P/\partial V)_T\, dV$ are called *partial differentials*. The combination of the partial differentials yields the *total differential* of the function. In general, then, if a variable $u = f(x_1, x_2, x_3, \ldots)$, where $x_1, x_2, x_3, \ldots$ are independent variables,[1] it follows that

$$du = \left(\frac{\partial u}{\partial x_1}\right)_{x_2, x_3, \ldots} dx_1 + \left(\frac{\partial u}{\partial x_2}\right)_{x_1, x_3, \ldots} dx_2 + \left(\frac{\partial u}{\partial x_3}\right)_{x_1, x_2, \ldots} dx_3 + \cdots \tag{4-16}$$

To illustrate the physical significance of Equation (4-16), consider the following example: The volume of a cylinder is a function of both the radius r of the cylinder and the height h of the cylinder and is given by the equation

$$V = f(r, h) = \pi r^2 h \tag{4-17}$$

Any change in either r or h will thus result in a change in V. The total differential of V, then, is

$$dV = \left(\frac{\partial V}{\partial r}\right)_h dr + \left(\frac{\partial V}{\partial h}\right)_r dh \tag{4-18}$$

Let us examine what each term in the expression means physically. For each incremental change dr in r, or dh in h, the volume changes. However, the manner in which the volume changes with r is different from the manner in which it changes with h. We see this by differentiating Equation (4-17) partially to yield

$$\left(\frac{\partial V}{\partial r}\right)_h = 2\pi r h \quad \text{and} \quad \left(\frac{\partial V}{\partial h}\right)_r = \pi r^2$$

[1]We find that Equation (4-16) will still be valid even if not all of the variables are independent.

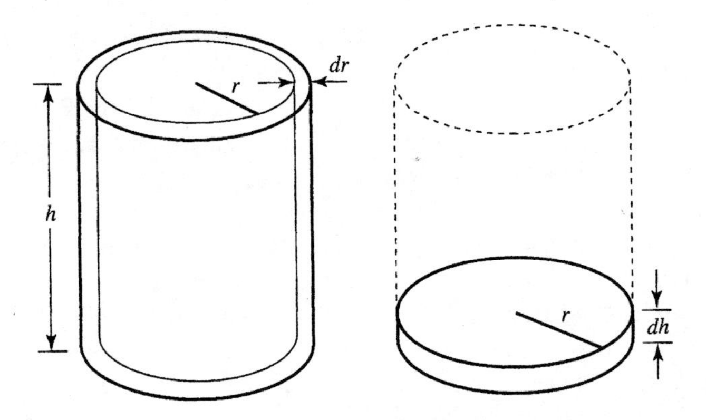

Figure 4-1 Partial differentials of a cylinder.

Hence,

$$dV = 2\pi rh\, dr + \pi r^2\, dh \tag{4-19}$$

We see, then, that there are at least two ways to consider the volume of the cylinder and changes in that volume. On the one hand, the quantity $2\pi rh\, dr$ is the volume of a hollow cylinder of thickness *dr*, shown in Fig. 4-1. The total volume of the cylinder can be thought of as summing together concentric cylinders of volume $2\pi rh\, dr$ until the radius *r* is reached. Hence, any change in the radius of the cylinder will affect the volume by adding or subtracting concentric cylinders.

On the other hand, the quantity $\pi r^2\, dh$ represents the volume of a thin disk of thickness *dh*, also shown in Fig. 4-1. Thus, in this case, the total volume can be thought of as summing together these thin disks until the height *h* is reached. Any change in the height will change the volume by adding or subtracting thin disks. The sum of these two effects results in the total change in the volume of the cylinder.

4-5 DERIVATIVE AS A RATIO OF INFINITESIMALLY SMALL CHANGES

In Section 4-1, we defined the derivative of $y = f(x)$ with respect to *x* as the ratio of the change in *y* to the change in *x* as the change in *x* becomes vanishingly small. We might ask, at this point, why such small changes are so important to the study of physical chemistry? After all, we never observe these small changes in our everyday experience. What, then, is the physical significance of the derivative?

To help answer these questions, consider the following example: The energy of a system is known to be a function of the temperature of the system; that is, $E = f(T)$. We saw in Chapter 2 that when we graph variables, such as *E* versus *T*, the relationship between changes in the two variables at some point (*T*, *E*) on the curve is given by the slope of a line drawn tangent to the curve at that point. This relationship on a curve of E versus T is called the *heat capacity* of the system and is denoted by the symbol c_V. The subscript V is necessary, because we find that E also is a function of the volume V of the system, and for our discussion here, we are taking V to be constant. Thus, the heat capacity at constant volume, c_V, is the slope of the tangent line drawn to the curve of E versus T at the point (T, E).

Consider, now, some finite change in energy, $\Delta E = E_2 - E_1$, with respect to a finite change in temperature, $\Delta T = T_2 - T_1$. A little experience shows us that the change in energy with respect to the change in temperature represents the slope of the curve only when the relationship between E and T is linear, as shown in Fig. 4-2(a). Under these circumstances, it is necessary that c_V be constant with temperature. We find experimentally, however, that c_V is rarely constant with temperature, and therefore $\Delta E/\Delta T$ is a poor approximation to the slope of the curve when E does not vary linearly with temperature, as shown in Fig. 4-2(b). Note that the ratio $\Delta E/\Delta T$, given by the line $\overline{ab}$, is quite different from the tangent to the curve at point a, designated as

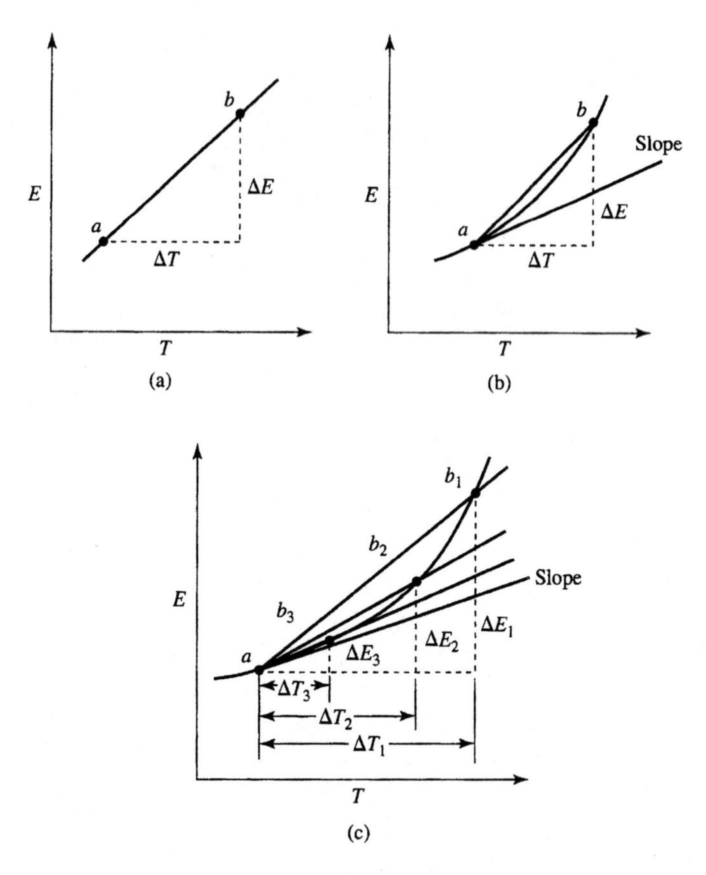

Figure 4-2 Internal energy as a function of temperature.

the slope. We see, though, that as we allow ΔT to become smaller and smaller, the ratio $\Delta E/\Delta T$, given by lines $\overline{ab_1}$, $\overline{ab_2}$, $\overline{ab_3}$, etc., approaches the slope of the curve at point a more and more closely, as illustrated in Fig. 4-2(c). In fact, in the limit as ΔT goes to zero, the ratio $\Delta E/\Delta T$ equals the slope of the curve exactly at point a. But this is just the definition of the derivative; that is,

$$\lim_{\Delta T \to 0} \frac{\Delta E}{\Delta T} = \frac{dE}{dT} = \text{slope of the curve}$$

Thus, one useful property of the derivative is that it represents the slope of a curve (actually, the slope of a line drawn tangent to the curve) at any point along the curve.

Another example of the importance in physical chemistry of infinitesimally small changes is found in the concept of *reversibility*. A reversible process is a process that, after taking place, can be reversed, restoring the system to exactly the state it was in before the process took place. To be reversible, the process must take place along a path in which all intermediate states are equilibrium states. Such a process must be defined as one in which the driving force at each step along the process is only infinitesimally larger than the opposing force. If we do not define the process in this way, the intermediate states will not be equilibrium states and the process will not be reversible. To illustrate this point, suppose that we have a cylinder fitted with a frictionless piston holding 1 mole of an ideal gas at some initial pressure, volume, and temperature, P_1, V_1, and T, as shown in Fig. 4-3. Suppose also that both the gas pressure and the external pressure on the gas are initially 1 bar. Next, let us assume that the piston is pinned into place and the external pressure is dropped to 0.5 bar. Then, when the pin is removed, the gas will expand suddenly, pushing the piston out to a new volume V_2, until the gas pressure is 0.5 bar (or the piston comes into contact with a new pin). In doing so, the gas will do a certain amount of work $w = -P_{\text{ext}}\,\Delta V = -0.5\,\Delta V$ on the surroundings. If we assume that the process takes place isothermally (i.e., at constant temperature), then the gas must absorb an amount of heat energy equal to $+0.5\,\Delta V$ from the surroundings.

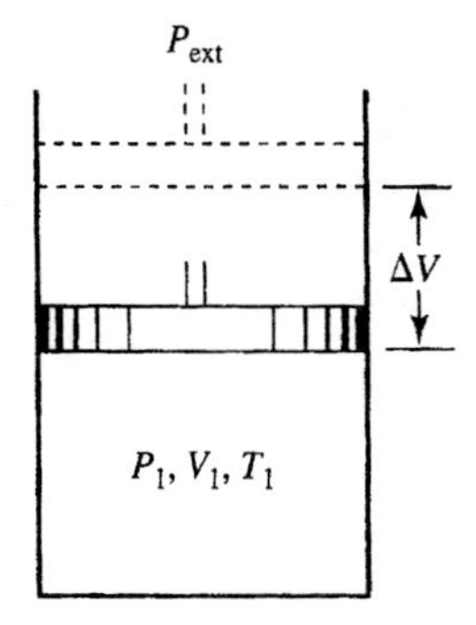

Figure 4-3 System consisting of a gas in a cylinder closed by a weightless piston.

The gas is now in a state characterized by P_2, V_2, and T. If the process just described were reversible, then it would take exactly the same amount of work to compress the gas back to its original volume V_1. The problem, however, is that, in order to compress the gas back to its original volume, at which the pressure of the gas is 1 bar, we need to use an external pressure of at least 1 bar. Thus, it will require twice the work $w = -1.0\ \Delta V$ to compress the gas back to V_1. Moreover, to keep the temperature constant, the gas must release to the surroundings twice the heat energy it absorbed in the expansion. The process is thus not reversible.

Let us now repeat the preceding expansion, but this time assuming that the external pressure on the gas at every point during the expansion is only infinitesimally smaller than the gas pressure. That is,

$$P_{\text{ext}} = P_{\text{gas}} - dP$$

For all practical purposes, we can now assume that, throughout the expansion, the external pressure equals the gas pressure. As the gas expands from volume V_1 to V_2, the gas pressure will drop from 1 bar to 0.5 bar, as it did before. The external pressure, however, also will do the same at every point during the expansion. Moreover, since the external pressure and the gas pressure differ by only an infinitesimal amount, the expansion should take place in an infinite amount of time, allowing equilibrium to be established at each point during the expansion. Again, the gas will do work on the surroundings, and to remain at constant temperature, the gas must absorb from the surroundings an amount of heat energy equal to the work done. To reverse the process in this case, we now compress the gas back to its original volume by making the external pressure only infinitesimally larger than the gas pressure at each point during the compression. That is,

$$P_{\text{ext}} = P_{\text{gas}} + dP$$

Because the external pressure at each point during the compression differs only infinitesimally from the external pressure at that point during the expansion (essentially, they both equal P_{gas}), the work done in the compression will be exactly equal, but opposite, to the work done by the gas during the expansion. The gas will release exactly the same amount of heat energy to the surroundings as it absorbed in the expansion, keeping the system isothermal. Thus, by employing infinitesimally small changes throughout the process, each intermediate step is allowed to reach equilibrium, and the process is reversible.

Examples

(a) One mole of an ideal gas at 300 K is allowed to expand isothermally against a constant external pressure of 1.00 bar from a volume of 1.00 liter to a volume of 5.00 liters. Find the work done by the gas and the heat energy absorbed from the surroundings.

Solution. For a constant external pressure,

$$w = -P_{ext}(V_2 - V_1) = -1.00(5.00 - 1.00)$$
$$= -4.00\ \ell \cdot \text{bar} = -400\ \text{J}$$

Since the process is isothermal and involves an ideal gas, $\Delta E = 0$ and

$$q = -w = +400\ \text{J}$$

(b) Find the work necessary to compress the gas in Example (a) back to its original volume, using a constant external pressure. Is the process in (a) reversible?

Solution. Before the expansion takes place, the pressure of the gas is

$$P_1 = \frac{nRT}{V_1} = \frac{(1.00)(0.08314)(300)}{(1.00)} = 24.9\ \text{bar}$$

Therefore, to compress the gas back to this pressure, we must use an external pressure at least equal to 24.9 bar. Thus, we have

$$w = -P_{ext}(V_2 - V_1) = -24.9(1.00 - 5.00) = +99.6\ \ell \cdot \text{bar} = 9{,}960\ \text{J}$$

and the process is not reversible.

(c) Find the work done by the gas if the process in (a) takes place reversibly.

Solution. For a reversible isothermal process involving an ideal gas, $w = -nRT \ln \frac{V_2}{V_1}$, so

$$w = -(1.00)(8.314)(300)\ln\frac{5.00}{1.00} = -4014\ \text{J}$$

Note that the equation describing this reversible isothermal process involves only the initial and final volumes. The work done compressing the gas back to V_1 is +4014 J.

We shall revisit the concept of reversibility in the next chapter, when we consider how the equation used to calculate the reversible work in Example (c) relates to the sum of infinitesimally small changes.

4-6 GEOMETRIC PROPERTIES OF DERIVATIVES

In the previous section, we introduced the idea that in the limiting case the derivative represents an instantaneous rate of change of two variables. Hence, if, for example, $y = f(x)$ is plotted on a two-dimensional Cartesian coordinate system, then dy/dx is

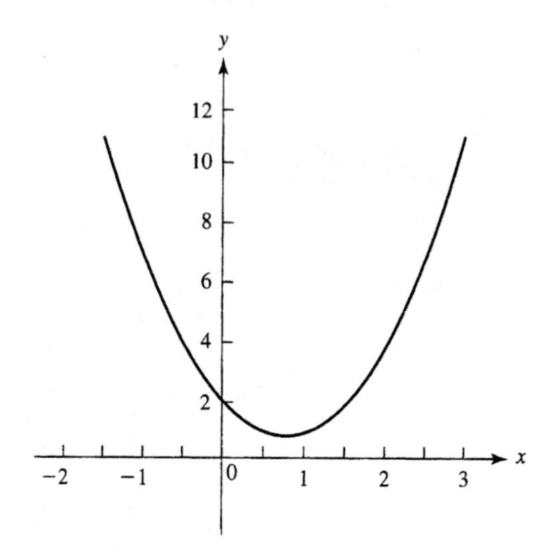

Figure 4-4 Graph of the function $y = 2x^2 - 3x + 2$.

the slope of the curve at any point (x, y) on the curve. With the exception of the function $y(x)$ = constant, functions either increase or decrease as the value of x increases. By looking at the derivative (or slope) evaluated at the point (x, y), we can determine whether the function $f(x)$ is increasing or decreasing as x increases without actually having to graph the function. If dy/dx is positive, then $f(x)$ increases as x increases; if dy/dx is negative, then $f(x)$ decreases as x increases.

Certain functions, such as parabolas (Fig. 4-4), or functions of higher order, such as cubic functions (Fig. 4-5), have either maximum or minimum values or both. Differential calculus can be used to help us determine the point or points along a curve at which maxima or minima occur. Since the slope of the curve must equal zero at these points, the first derivative also must be zero. For example, the parabola shown in Fig. 4-4 is described by the equation

$$y = 2x^2 - 3x + 2$$

Taking the first derivative gives

$$\frac{dy}{dx} = 4x - 3$$

Setting the first derivative to zero and solving for x, we have

$$4x - 3 = 0 \quad \text{or} \quad x = \frac{3}{4}$$

Substituting this value of x into the equation yields $y = 0.875$, which gives the minimum point on the curve. To determine whether the function is a maximum or a

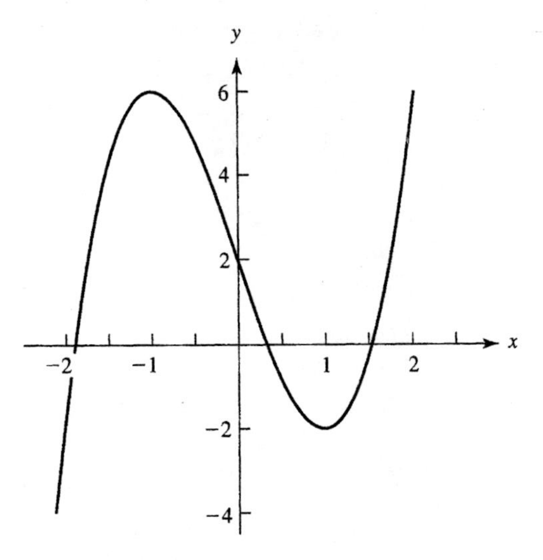

Figure 4-5 Graph of $y = 2x^3 - 6x + 2$.

minimum at this point without actually having to graph the curve, we can substitute into the equation for the curve both values for x that are greater than the value at the point where the derivative is zero and values for x that are smaller than the value at the point where the derivative is zero. Then we note the behavior of y. A simpler way to test whether the function is a maximum or a minimum is to look at the *second* derivative of the function, evaluated at the point of zero slope:

If $\dfrac{d^2y}{dx^2} < 0$, then the function is a maximum.

If $\dfrac{d^2y}{dx^2} > 0$, then the function is a minimum.

If $\dfrac{d^2y}{dx^2} = 0$, then the function is at a *point of inflection*—a point on the curve where the curve changes from one that exhibits a maximum at some point (i.e., is "concave downward") to one that that exhibits a minimum at some point (i.e., is "concave upward"), or vice versa.

Now, consider the cubic function shown in Fig. 4-5:

$$y = 2x^3 - 6x + 2$$

Taking the first derivative of this equation and setting it equal to zero gives

$$\frac{dy}{dx} = 6x^2 - 6 = 0; x^2 - 1 = 0; x = \pm 1$$

which indicates that there is a maximum or a minimum at $x = +1$ and $x = -1$. Taking the second derivative of the equation gives

$$\frac{d^2y}{dx^2} = 12x$$

For $x = +1$, $d^2y/dx^2 = 12$, which indicates that the function reaches minimum at that value of x. For $x = -1$, $d^2y/dx^2 = -12$, which indicates that the function reaches a maximum at *that* value of x. Note that a point of inflection occurs at $x = 0$.

Examples

1. The total volume, in milliliters, of a glucose–water solution is given by the equation

$$V = 1001.93 + 111.5282m + 0.64698m^2$$

where m is the molality of the solution. The partial molar volume of glucose, $\overline{V}_{\text{glucose}}$, is the slope $\partial V/\partial m$ of a curve of V versus m. Find the partial molar volume of glucose in a 0.100 m solution of glucose in water.

Solution. Taking the derivative of V versus m gives

$$\overline{V}_{\text{glucose}} = \frac{\partial V}{\partial m} = 111.5282 + 1.2940m$$

Substituting the concentration $m = 0.100$ into this equation results in

$$\overline{V}_{\text{glucose}} = 111.6576 \text{ ml}$$

2. The probability of a gas molecule having a speed c lying in the range between c and $c + dc$ is given by the Maxwell distribution law

$$P_c dc = 4\pi\left(\frac{m}{2\pi kT}\right)^{3/2} e^{-mc^2/2kT} c^2 dc$$

where m, k, and T are constants. Find an expression for the most probable speed of the molecule.

Solution. The most probable speed c_{mp} occurs at the point where the probability distribution function P_c is a maximum. Thus, we must maximize the function

$$P_c = 4\pi\left(\frac{m}{2\pi kT}\right)^{3/2} e^{-mc^2/2kT} c^2$$

with respect to c. Taking the derivative of P_c with respect to c and setting it equal to zero gives

$$\frac{dP_c}{dc} = 4\pi\left(\frac{m}{2\pi kT}\right)^{3/2}\left[e^{-mc^2/2kT}(2c) + c^2\left(-\frac{2mc}{2kT}\right)e^{-mc^2/2kT}\right] = 0$$

We can divide through this equation by $4\pi(m/2\pi kT)^{3/2}(2c)e^{-mc^2/2kT}$, which leaves

$$1 - \frac{mc^2}{2kT} = 0$$

$$c_{\text{mp}} = \sqrt{\frac{2kT}{m}}$$

3. In the consecutive reaction $A \rightarrow B \rightarrow C$, the molar concentration of B follows the first-order rate law given by the equation

$$(B) = \frac{(A)_0 k_1}{k_2 - k_1}[e^{-k_1 t} - e^{-k_2 t}]$$

where $(A)_0$ is the initial concentration of A and k_1 and k_2 are specific rate constants. Find the value of t for which (B) is a maximum.

Solution. Taking the derivative of (B) with respect to t and setting it equal to zero gives

$$\frac{d(B)}{dt} = \frac{(A)_0 k_1}{k_2 - k_1}[-k_1 e^{-k_1 t} + k_2 e^{-k_2 t}] = 0$$

Dividing through by $(A)_0 k_1/(k_2 - k_1)$ results in

$$k_2 e^{-k_2 t} = k_1 e^{-k_1 t}$$

Taking the natural logarithm of this equation, we have

$$\ln k_2 - k_2 t_{\max} = \ln k_1 - k_1 t_{\max}$$

$$t_{\max} = \frac{\ln(k_2/k_1)}{(k_2 - k_1)}$$

4-7 CONSTRAINED MAXIMA AND MINIMA

There are a number of problems in physical chemistry in which it is necessary to maximize or minimize a function under specific restrictive conditions. For example, suppose we wish to maximize some function $f(x, y)$, subject to the restriction that another function of x and y, $\phi(x, y)$, always equals zero. We can do this by a method known as *Lagrange's method of undetermined multipliers*. In order to maximize $f(x, y)$ by the Lagrange method, consider the total differentials

$$df = \left(\frac{\partial f}{\partial x}\right)_y dx + \left(\frac{\partial f}{\partial y}\right)_x dy \tag{4-20}$$

and

$$d\phi = \left(\frac{\partial \phi}{\partial x}\right)_y dx + \left(\frac{\partial \phi}{\partial y}\right)_x dy = 0 \tag{4-21}$$

(Since $\phi(x, y) = 0, d\phi = 0$). Equations (4-20) and (4-21) can now be combined by solving Equation (4-21) for dy and substituting the result back into Equation (4-20). That is,

$$dy = -\frac{\left(\frac{\partial \phi}{\partial x}\right)}{\left(\frac{\partial \phi}{\partial y}\right)} dx \tag{4-22}$$

and

$$df = \left[\frac{\partial f}{\partial x} - \frac{\partial f}{\partial y} \frac{\left(\frac{\partial \phi}{\partial x}\right)}{\left(\frac{\partial \phi}{\partial y}\right)}\right] dx \tag{4-23}$$

Note that this procedure effectively removes the explicit y-dependence in Equation (4-23); hence, the function f can now be treated as a function of a single variable x. Thus, the function reaches a maximum at the point where $df/dx = 0$. This gives

$$\left[\frac{\partial f}{\partial x} - \frac{\partial f}{\partial y} \frac{\left(\frac{\partial \phi}{\partial x}\right)}{\left(\frac{\partial \phi}{\partial y}\right)}\right] = 0$$

or

$$\frac{\left(\frac{\partial f}{\partial x}\right)}{\left(\frac{\partial f}{\partial y}\right)} = \frac{\left(\frac{\partial \phi}{\partial x}\right)}{\left(\frac{\partial \phi}{\partial y}\right)} \equiv \lambda \tag{4-24}$$

where λ is a constant called an *undetermined multiplier*. Rearranging Equation (4-24), we obtain the two equations

$$\frac{\partial f}{\partial x} - \lambda \frac{\partial \phi}{\partial x} = 0 \quad \text{and} \quad \frac{\partial f}{\partial y} - \lambda \frac{\partial \phi}{\partial y} = 0$$

or

$$\frac{\partial}{\partial x}(f - \lambda\phi) = 0 \quad \text{and} \quad \frac{\partial}{\partial y}(f - \lambda\phi) = 0 \tag{4-25}$$

which, along with $\phi(x, y) = 0$, allows us to determine both the point at which a maximum occurs and λ.

We find from experience that the extension of Lagrange's method to include more than one restriction requires that there be at least one more independent variable than there are restrictions. Thus, for more than one restriction, we have

$$\begin{aligned} F(x, y, z, \ldots) = f(x, y, z, \ldots) &- \alpha u(x, y, z, \ldots) \\ &- \beta v(x, y, z, \ldots) - \cdots \end{aligned} \tag{4-26}$$

where $f(x, y, z, \ldots)$ is the function to be maximized, $u(x, y, z, \ldots)$ and $v(x, y, z, \ldots)\ldots$ are the restrictions, and α, β, $\ldots$ are the undetermined multipliers. The condition for constrained maximization or minimization of $f(x, y, z, \ldots)$ is

$$\left(\frac{\partial F}{\partial x}\right) = 0, \left(\frac{\partial F}{\partial y}\right) = 0, \left(\frac{\partial F}{\partial z}\right) = 0, \ldots \tag{4-27}$$

Examples

1. Find the dimensions of a rectangular area for which the area is a maximum, but the circumference is a minimum.

Solution. The area of a rectangle is $A = ab$, where a and b are the sides. The circumference of that same rectangle is $C = 2(a + b)$. We wish to maximize A while minimizing C. Let $\phi = C - 2(a + b) = 0$. Therefore, Equation (4-26) for this problem is

$$F(a, b) = ab - \lambda\phi = ab - \lambda C + 2\lambda(a + b)$$

The condition for the maximization of A is that

$$\left(\frac{\partial F}{\partial a}\right) = 0 \quad \text{and} \quad \left(\frac{\partial F}{\partial b}\right) = 0$$

Taking partial derivatives gives

$$\frac{\partial F}{\partial a} = b - \lambda\frac{\partial C}{\partial a} + 2\lambda = 0 \quad \text{and} \quad \frac{\partial F}{\partial b} = a - \lambda\frac{\partial C}{\partial b} + 2\lambda = 0$$

Since C is a minimum, it follows that $\partial C/\partial a = \partial C/\partial b = 0$. Therefore, we can write

$$b + 2\lambda = 0 \quad \text{and} \quad a + 2\lambda = 0$$

or

$$a = b$$

The rectangular shape with the maximum area and minimum circumference is thus a square.

2. A problem in statistical mechanics requires maximizing the function

$$f(n_0, n_1, n_2, \ldots) = n \ln n - \sum_{n_i=0}^{n} n_i \ln n_i$$

subject to the conditions that

$$\sum n_i = n \quad \text{and} \quad \sum n_i E_i = E$$

Solution. Let

$$u(n_0, n_1, n_2, \ldots) = \sum n_i - n = 0$$

and

$$v(n_0, n_1, n_2, \ldots) = \sum n_i E_i - E = 0$$

Then

$$F(n_0, n_1, n_2, \ldots) = n \ln n - n - \sum_{n_i=0}^{n} n_i \ln n_i - \alpha \sum (n_i - n)$$
$$- \beta \sum (n_i E_i - E)$$

where α and β are undetermined multipliers. The condition for the constrained maximization of $n \ln n - \sum_{n_i=0}^{n} n_i \ln n_i$ is that

$$\left(\frac{\partial F}{\partial n_0}\right) = 0, \quad \left(\frac{\partial F}{\partial n_1}\right) = 0, \quad \text{and} \quad \left(\frac{\partial F}{\partial n_2}\right) = 0, \ldots \quad \text{or}$$
$$\frac{\partial F}{\partial n_j} = 0; \quad j = 0, 1, 2, 3, \ldots$$

Taking the derivative of F with respect to each n_j in the sum, and recalling that $n_j \ln n_j$ must be differentiated as a product and that n and E are constants, we obtain

$$\frac{\partial F}{\partial n_j} = -n_j\left(\frac{1}{n_j}\right) - \ln n_j - \alpha - \beta E_j = 0$$

so that

$$\ln n_j = -1 - \alpha - \beta E_j$$

Taking the antilogarithm of this equation gives

$$n_j = e^{-(1+\alpha)}e^{-\beta E_j}$$

But $n = \sum n_j$; therefore, $n = e^{-(1+\alpha)}\sum e^{-\beta E_j}$, which gives the Boltzmann distribution equation

$$\frac{n_j}{n} = \frac{e^{-\beta E_j}}{\sum e^{-\beta E_j}}$$

The function $n \ln n - \sum_{n_i-0}^{n} n_i \ln n_i$ will be a maximum when

$$\frac{n_0}{n} = \frac{e^{-\beta E_0}}{q_s}; \frac{n_1}{n} = \frac{e^{-\beta E_1}}{q_s}; \frac{n_2}{n} = \frac{e^{-\beta E_2}}{q_s}; \ldots$$

where $q_s = \sum e^{-\beta E_j}$ is known as the *system partition function.*

SUGGESTED READINGS

1. BRADLEY, GERALD L., and SMITH, KARL J. *Calculus*. Upper Saddle River, NJ: Prentice Hall, 1995.
2. VARBERG, DALE, and PURCELL, EDWIN J. *Calculus*, 7th ed. Upper Saddle River, NJ: Prentice Hall, 1997.

PROBLEMS

1. Differentiate the following functions, assuming that lowercase letters are variables and uppercase letters are constants:

(a) $y = 3x^3 + 4x^2 - 5x + 6$

(b) $y = \sqrt{1 - x^2}$

(c) $y = x^2 - 9x + 16$

(k) $w = N \ln N - n_i \ln n_i$

(l) $s = \ln t \cdot e^{-3t}$

(m) $\ln g = \frac{A}{t} + t \ln t$

(d) $r = 5 \tan 2\theta$

(e) $y = x^4 e^{3x}$

(f) $y = A \sin\theta \cos\theta$

(g) $y = x^5\sqrt{1 - e^{2x}}$

(h) $y = x^{-2}(1 - e^x)\sin 3x$

(i) $y = \dfrac{x^2}{\sqrt{1 - 5x}}$

(j) $y = \ln(1 - e^x)$

(n) $e = \dfrac{E^2}{A}\left(z^2 - \dfrac{27}{8}z\right)$

(o) $\phi = 2A\cos\left(\dfrac{N\pi x}{L}\right)$

(p) $\ln p = \dfrac{-\Delta H}{Rt} + K$

(q) $\ln k = -\dfrac{\Delta G}{Rt}$

(r) $u = \dfrac{A}{r^{12}} - \dfrac{B}{r^6}$

(s) $d = \dfrac{M}{v}$

(t) $\phi = Ae^{-B/Rt}$

2. Evaluate the following partial derivatives:

(a) $PV = nRT$; P with respect to V

(b) $\left(P + \dfrac{n^2a}{V^2}\right)(V - nb) = nRT$; P with respect to V

(c) $\rho = \dfrac{PM}{RT}$; ρ with respect to T

(d) $H = a + bT + cT^2 + \dfrac{d}{T}$; H with respect to T

(e) $r = \sqrt{(x^2 + y^2 + z^2)}$; r with respect to y.

(f) $y = r\sin\theta\cos\phi$; y with respect to ϕ

(g) $\left(\dfrac{\partial S}{\partial T}\right)_P = \dfrac{1}{T}\left(\dfrac{\partial H}{\partial T}\right)_P$; S with respect to P at constant T

(h) $\left(\dfrac{\partial S}{\partial P}\right)_T = \dfrac{1}{T}\left[\left(\dfrac{\partial H}{\partial P}\right)_T - V\right]$; S with respect to T at constant P

Note that $(\partial H/\partial P)$ also is a function of T.

(i) $D = \sin\theta\sin\phi\cos\phi$; D with respect to ϕ

(j) $E = \dfrac{c_A^2H_{AA} + c_B^2H_{BB} + 2c_Ac_BH_{AB}}{c_A^2 + c_B^2 + 2c_Ac_BS_{AB}}$; E with respect to c_B.

(k) $q = \sum e^{-E_i/kT}$; q with respect to E_i

(l) $q = \sum e^{-E_i/kT}$; q with respect to T

3. Determine the slope of each of the following curves at the indicated points:

(a) $y = x^3$ at $x = 5$

(b) $y = 2x^3 - 5x^2 + 4x - 6$ at $x = 2$

(c) $y = 3 \ln 3x$ at $x = 4$

(d) $y = x \ln 2x$ at $x = 5$

(e) $r = 10 \cos \theta$ at $\theta = \pi$

(f) $r = 5 \sin \theta \cos \theta$ at $\theta = \pi/2$

(g) $y = (x^2 - 6)^{1/2}$ at $x = 4$

(h) $s = \frac{1}{2}At^2$ at $t = 30$ seconds, where acceleration $A = 9.80 \text{ m/s}^2$ is constant.

(i) $C_P = 25.90 + 33.00 \times 10^{-3}T - 30.4 \times 10^{-7}T^2$ at $T = 298.2$ K

(j) $\ln P = -\Delta H/RT + B$ at $T = 298.2$ K, where $\Delta H = 30{,}820$ J/mol, $R =$ 8.314 J/mol·K, and $B = 2.83$ are constants.

(k) $(A) = (A)_0 e^{-kt}$ at $t = 5.0$ hours, where $(A)_0 = 0.0200\ M$ and $k = 4.11 \times 10^{-2} \text{ hr}^{-1}$ are constants.

4. For each of the functions that follow, determine whether it reaches a maximum value, a minimum value, or both. Evaluate each function that does reach a maximum or minimum value at any such point. Specify any points of inflection.

(a) $y = 3x^2 - 5x + 2$

(b) $y = 2x^3 - 3x^2 + 5x - 12$

(c) $y = \sin 2x$

(d) $\psi = Ae^{mx}$, where A and m are constants.

(e) $U(r) = 4e\left[\left(\frac{\sigma}{r}\right)^{12} - \left(\frac{\sigma}{r}\right)^{6}\right]$, where e and σ are constants.

(f) $\psi = \frac{1}{2}(1 + \sin \theta) + \sqrt{2} \cos \theta$

(g) $E = \frac{e^2}{a}\left(z^2 - \frac{27}{8}z\right)$, where e and a are constants.

(h) $P_E = 2\pi^{-1/2}(kT)^{-3/2}e^{-E/kT}E^{1/2}$, where π, k, and T are constants.

(i) $P(x) = \left(\frac{2}{a}\right)\sin^2\frac{\pi x}{a}$ between $x = 0$ and $x = a$

(j) $U(r) = -N_0 A\frac{z^2}{r} + \frac{B}{r^n}$, where N_0, A, z, and B are constants.

5. The rate constant for a chemical reaction is found to vary with temperature according to the Arrhenius equation

$$k = Ae^{-E_a/RT}$$

where A, E_a, and R are constants. Find an expression that describes the change in k with respect to T.

6. The density of an ideal gas is found to vary with respect to temperature according to the equation

$$\rho = \frac{PM}{RT}$$

where P, M, and R are considered to be constant in this case. Find an expression that describes the slope of the curve of ρ versus T.

7. Find the partial derivative of P with respect to T for a gas obeying van der Waals equation

$$\left(P + \frac{n^2a}{V^2}\right)(V - nb) = nRT$$

8. Find the partial derivative of P with respect to V for the gas in Problem 7.
9. A certain gas obeys the equation of state

$$P(V - nb) = nRT$$

where n and R are constants in this case. Determine the coefficient of expansion,

$$\alpha = \left(\frac{1}{V}\right)\left(\frac{\partial V}{\partial T}\right)_P$$

of this gas.
10. The volume of an ideal gas is simultaneously a function of the pressure and the temperature of the gas. Write an equation for the total differential of V. Using the ideal gas law for 1 mole of gas, $PV = RT$, evaluate the partial derivatives in that equation.
11. The vibrational potential energy of a diatomic molecule can be approximated by the Morse function

$$U(r) = A(1 - e^{-B(r-r_0)})^2$$

where A, B, and r_0 are constants. Find the value of r for which U is a minimum.
12. The equation describing the realm of spatial possibilities for a particle confined in a one-dimensional "box" in the state $n = 1$ is

$$\psi(x) = \sqrt{\frac{2}{a}}\sin\frac{\pi x}{a}$$

where a is the length of the box. Find the value of x for which $\psi(x)$ is a maximum.
13. The distribution function describing the probability of finding an electron in the $1s$ orbital of the hydrogen atom a certain distance from the nucleus, irrespective of direction, is given by the equation

$$P(r) = 4\left(\frac{1}{a_0}\right)^3 e^{-2r/a_0}r^2$$

Show that the function reaches a maximum at the point where $r = a_0$, the Bohr radius.
14. The van der Waals equation

$$\left(P + \frac{n^2a}{V^2}\right)(V - nb) = nRT$$

is a cubic equation in volume. Consequently, the equation cannot be solved for $V = f(T, P)$. Find an expression for the partial derivative $(\partial V/\partial T)_P$ for a van der Waals gas.

CHAPTER 5 Integral Calculus

5-1 INTRODUCTION

There are basically two major approaches to integral calculus. One is to consider the integral as an antiderivative. In this approach, *integration*, the process of taking integrals, is considered the inverse of differentiation. The other approach is to consider the integral as the sum of many similar, infinitesimal elements. The integral then has a value that is equal to the area under a curve. The second approach allows us to assign a physical meaning to the integral. Introductory courses on integral calculus spend a tremendous amount of time on the first approach, teaching all the various methods for generating integrals. While this is, no doubt, important and should be learned at some point, in practice it is rarely used, since most of us learn to refer to tables of integrals to do our integrating. However, some general and special methods of integration are reviewed in this chapter, primarily because some functions are not always in the form found in tables of integrals.[1]

In previous chapters, we studied the mathematics associated with dividing a function into many small, incremental parts and determining the effects of the incremental changes on the variables of the function. In this chapter, we shall consider the reverse process: Knowing the effect of the individual changes, we wish to determine the overall effect of adding together these changes such that the sum equals a finite change. Before considering the physical significance and the applications of integral calculus to physical chemistry, let us review the general and special methods of integration.

5-2 INTEGRAL AS AN ANTIDERIVATIVE

In Chapter 4, we considered the differentiation of the function $y = f(x)$:

$$\frac{dy}{dx} = \frac{df(x)}{dx} = f'(x) \tag{5-1}$$

[1]A complete table of integrals is given in Appendix II. Also, excellent tables of integrals are found in any edition of the *CRC Handbook of Chemistry and Physics*, by David R. Lide (Boca Raton, FL: CRC Press).

In differential form,

$$dy = f'(x)\,dx \tag{5-2}$$

where $f'(x)$ denotes the first derivative of the function $f(x)$ with respect to x. In this section, we pose the following question: What function $f(x)$, when differentiated, yields the function $f'(x)$? For example, we might ask what function $f(x)$, when differentiated, yields the function $f'(x) = 2x$? Substituting $f'(x) = 2x$ into Equation (5-2) gives

$$dy = 2x\,dx \quad \text{or} \quad \frac{dy}{dx} = 2x$$

The function $f(x)$ which we are seeking is called the *integral* of the differential and is symbolized by the equation

$$f(x) = \int f'(x)\,dx \tag{5-3}$$

where the symbol $\int$ is called the *integral sign.*[2] The function that is to be integrated, $f'(x)$, is called the *integrand.*

In this case, it is not too difficult to see by inspection that if $f'(x) = 2x$, then $f(x) = x^2$, since, if one differentiates x^2, one obtains the derivative $f'(x) = 2x$. The term $f(x) = x^2$ is not the complete solution, however, because differentiation of the function $f(x) = x^2 + C$, where C is a constant, also will give $f'(x) = 2x$. Hence, it is possible that the integral may contain a constant, called the *constant of integration*, and this constant *always* is included as part of the answer to any integration problem. Thus,

$$y = \int 2x\,dx = x^2 + C$$

5-3 GENERAL METHODS OF INTEGRATION

Let us now consider several general methods of integration. The following list gives the standard integrals for most of the functions that are important to physical chemistry (for a complete table of integrals, see Appendix II):

1. $\int du(x) = u(x) + C$
   The integral of the differential of a function is equal to the function itself.
2. $\int a\,du = a\int du = au + C$
   Since a is a constant factor, it can be brought out of the integral sign and multiplied by the value of the integral after the integration is completed.
3. $\int u^n\,du = \dfrac{u^{n+1}}{n+1} + C$, where $n \neq -1$.

[2]The integral sign evolved from an elongated S that originally stood for "summation."

Examples:

(a) $\int x^3\,dx = \frac{x^4}{4} + C$

(b) $\int \frac{\Delta H}{RT^2} dT = \frac{\Delta H}{R} \int \frac{1}{T^2} dT = \frac{\Delta H}{R} \int T^{-2}\,dT = -\frac{\Delta H}{R} T^{-1} + C$

$= -\frac{\Delta H}{RT} + C$

4. $\int \frac{du}{u} = \int d \ln u = \ln u + C$. This is one of the most used integrals in physical chemistry; it describes the overall effect of summing fractional changes. (See Section 3-4.)

Examples:

(a) $\int \frac{dx}{3x} = \frac{1}{3} \int \frac{dx}{x} = \frac{1}{3} \ln x + C$

(b) $-\int nRT \frac{dV}{V} = -nRT \int \frac{dV}{V} = -nRT \ln V + C$

(c) $\int \frac{d(A)}{(A)} = -k \int dt$

$\ln(A) = -kt + C$. Both integrals will have a constant of integration. It is customary to combine them.

(d) $\int \frac{dP}{P} = \int \frac{\Delta H}{RT^2} dT$

$\ln P = -\frac{\Delta H}{RT} + C$ (the Clausius–Clapeyron equation)

5. $\int [f(x) + g(x)]\,dx = \int f(x)\,dx + \int g(x)\,dx$

The integral of a sum is the sum of the integrals.

Example:

$$\Delta H = \int C_P\,dT = \int (a + bT + cT^2)\,dT$$

$$= a \int dT + b \int T\,dT + c \int T^2\,dT = aT + \frac{b}{2}T^2 + \frac{c}{3}T^3 + C$$

6. $\int e^{mx}\,dx = \frac{1}{m} e^{mx} + C$

7. $\int \sin kx\,dx = -\frac{1}{k} \cos kx + C$

8. $\int \cos ax\,dx = \frac{1}{k} \sin kx + C$

5-4 SPECIAL METHODS OF INTEGRATION

Many of the functions encountered in physical chemistry are not in one of the general forms just given. For this reason, we include several special methods of integration.

Algebraic Substitution We find that certain mathematical functions can be transformed into one of the general forms in Section 5-3 or into one of the forms found in the table of integrals by some form of algebraic substitution.

Examples

(a) Evaluate $\int 7x(1 - x^2)^5\,dx$.

Let us attempt to transform this integral into the form $\int u^n\,du$. Let $u = (1 - x^2)$. Then $du = -2x\,dx$. Hence,

$$\int 7x(1 - x^2)^5\,dx = -\frac{7}{2}\int u^5\,du = -\frac{7}{12}u^6 + C = -\frac{7}{12}(1 - x^2)^6 + C$$

(b) Evaluate $\int e^{-\Delta E/kT}\left(\frac{\Delta E}{kT^2}\right)dT$.

Let $u = -\frac{\Delta E}{kT}$. Then $du = \frac{\Delta E}{kT^2}dT$. Thus,

$$\int e^{-\Delta E/kT}\left(\frac{\Delta E}{kT^2}\right)dT = \int e^u\,du = e^u + C = e^{-\Delta E/kT} + C$$

(c) Evaluate $\int \frac{dV}{(V - nb)}$.

Let us attempt to transform the integral into the form $\int \frac{du}{u}$.

Let $u = (V - nb)$. Then $du = dV$. It follows that

$$\int \frac{dV}{(V - nb)} = \int \frac{du}{u} = \ln u + C = \ln(V - nb) + C$$

(d) Evaluate $\int \sin\frac{2\pi x}{L}e^{-ikx}\,dx$, using the table of integrals found in Appendix II.

Let $a = -ik$ and $b = 2\pi/L$. The integral then becomes $\int e^{ax} \sin bx\, dx$, and we have

$$\int e^{ax} \sin bx\, dx = \frac{e^{ax}(a \sin bx - b \cos bx)}{(a^2 + b^2)} + C$$

$$= \frac{e^{-ikx}\left[-ik \sin \frac{2\pi x}{L} - \frac{2\pi}{L} \cos \frac{2\pi x}{L}\right]}{\frac{4\pi^2}{L^2} - k^2} + C$$

(e) Evaluate $\int 2 \sin^2 x \cos x\, dx$.

Let $u = \sin x$. Then $du = \cos x\, dx$. Hence,

$$\int 2 \sin^2 x \cos x\, dx = 2\int u^2\, du = \frac{2}{3}u^3 + C = \frac{2}{3}\sin^3 x + C$$

Trigonometric Transformation Many trigonometric functions can be transformed into a proper form for integration by making some form of trigonometric transformation with the aid of trigonometric identities. For example, to evaluate the integral $\int \sin^2 x\, dx$, we must use the identity

$$\sin^2 x = \frac{1}{2}(1 - \cos 2x)$$

Thus,

$$\int \sin^2 x\, dx = \int \frac{1}{2}(1 - \cos 2x)\, dx = \frac{1}{2}\int dx - \frac{1}{2}\int \cos 2x\, dx$$

Integrating each term separately gives

$$\int \sin^2 x\, dx = \frac{x}{2} - \frac{1}{4}\sin 2x + C$$

Again, the integration of integrals of this type is more practically done by using the table of integrals (Appendix II).

Example

Evaluate the integral $\int \cos^3 2x\, dx$.
This function can be integrated with Integral (88) from the table of integrals. Here, $a = 2$ and $b = 0$. We have

$$\int \cos^3(ax + b)\,dx = \frac{1}{a}\sin(ax + b) - \frac{1}{3a}\sin^3(ax + b) + C$$

$$\int \cos^3 2x\,dx = \frac{1}{2}\sin 2x - \frac{1}{6}\sin^3 2x + C$$

Partial Fractions Consider an integral of the type

$$\int \frac{dx}{(a - x)(b - x)}, \quad \text{where } a \text{ and } b \text{ are constants}$$

This type of integral can be transformed into simpler integrals by the method of partial fractions. Let $A = (a - x)$ and $B - (b - x)$. Then

$$\frac{1}{A} - \frac{1}{B} = \frac{B}{AB} - \frac{A}{AB} = \frac{(B - A)}{AB}$$

Therefore,

$$\frac{1}{AB} = \frac{1}{(B - A)}\left(\frac{1}{A} - \frac{1}{B}\right)$$

$$\frac{1}{(a - x)(b - x)} = \frac{1}{(b - a)}\left(\frac{1}{(a - x)} - \frac{1}{(b - x)}\right), a \neq b$$

$$\int \frac{dx}{(a - x)(b - x)} = \frac{1}{b - a}\left[\int \frac{dx}{(a - x)} - \int \frac{dx}{(b - x)}\right]$$

which can be integrated to give

$$\int \frac{dx}{(a - x)(b - x)} = \frac{1}{(b - a)}[-\ln(a - x) + \ln(b - x)] + C$$

$$= \frac{1}{(b - a)}\ln\frac{(b - x)}{(a - x)} + C$$

Integration by Parts If we differentiate a product, we get

$$d(uv) = u\,dv - v\,du$$

Rearranging this equation gives

$$u\,dv = d(uv) - v\,du$$

Integrating the preceding equation, we have

$$\int u\,dv = \int d(uv) - \int v\,du = uv - \int v\,du$$

Many functions can be integrated by putting them into this form, known as *integration by parts.*

Example

Evaluate $\int e^{ax} \sin bx\,dx$.

Let $u = \sin bx$. Therefore, $du = b \cos bx\,dx$. Let $dv = e^{ax}\,dx$. Then $v = \frac{1}{a}e^{ax}$, and we have

$$\int e^{ax} \sin bx\,dx = uv - \int v\,du = \frac{1}{a}e^{ax} \sin bx - \int \frac{1}{a}e^{ax}(b \cos bx)\,dx$$
$$= \frac{1}{a}e^{ax} \sin bx - \frac{b}{a}\int e^{ax} \cos bx\,dx$$

We must now evaluate the second integral, $\int e^{ax} \cos bx\,dx$, by parts.

Let $u = \cos bx$. Then $du = -b \sin bx\,dx$. Let $dv = e^{ax}\,dx$. Therefore, $v = \frac{1}{a}e^{ax}$, and we have

$$\int e^{ax} \cos bx\,dx = \frac{1}{a}e^{ax} \cos bx - \int \frac{1}{a}e^{ax}(-b \sin bx)\,dx$$
$$= \frac{1}{a}e^{ax} \cos bx + \frac{b}{a}\int e^{ax} \sin bx\,dx$$

Hence,

$$\int e^{ax} \sin bx\,dx = \frac{1}{a}e^{ax} \sin bx - \frac{b}{a}\left[\frac{1}{a}e^{ax} \cos bx + \frac{b}{a}\int e^{ax} \sin bx\,dx\right]$$
$$\int e^{ax} \sin bx\,dx = \frac{1}{a}e^{ax} \sin bx - \frac{b}{a^2}e^{ax} \cos bx - \frac{b^2}{a^2}\int e^{ax} \sin bx\,dx.$$

Multiplying the numerator and denominator of the term on the left side of the equation by a^2 and the numerator and the denominator of the first term on the right side of the equation by a, we have

$$\frac{a^2}{a^2}\int e^{ax}\sin bx\,dx = \frac{a}{a^2}e^{ax}\sin bx - \frac{b}{a^2}e^{ax}\cos bx - \frac{b^2}{a^2}\int e^{ax}\sin bx\,dx$$

Collecting terms (note that the a^2 terms in the denominator cancel out) yields

$$(a^2 + b^2)\int e^{ax}\sin bx\,dx = e^{ax}(a\sin bx - b\cos bx)$$

$$\int e^{ax}\sin bx\,dx = \frac{e^{ax}(a\sin bx - b\cos bx)}{(a^2 + b^2)} + C$$

5-5 THE INTEGRAL AS A SUMMATION OF INFINITESIMALLY SMALL ELEMENTS

In the previous sections, we treated integration as the purely mechanical operation of finding antiderivatives. Let us now turn to the physical aspect of integration in order to understand the physical importance of the integral.

Consider the expansion of an ideal gas from a volume V_1 to a volume V_2 against a constant external pressure. In order to determine the work done in the process, we must know the external pressure on the gas at each successive step in the expansion (or compression). If we plot the external pressure on the gas versus the volume of the gas, as shown in Fig. 5-1, we obtain a graph known as an *indicator diagram.* Note carefully that the plot shown in the indicator diagram is *not* a graph of P_{ext} as a function of V: *There is no necessary relationship between the external pressure on the gas and the volume of the gas.* We find from physics that the work done in PV expansions and compressions is the negative of the area under a curve of P_{ext} versus V.[3] In the case where the external pressure is constant, $w = -P_{ext}\Delta V$, which is the negative of the area shown in the figure.

Consider, next, the more complicated case in which the external pressure changes for some reason during the course of the expansion, as shown on the indicator diagram in Fig. 5-2. Again, we emphasize that the external pressure is not necessarily changing as a function of volume.[4] The work done in this case is still the area under the curve of P_{ext} versus V. Measuring this area, however, is much more difficult than it was when the external pressure was constant.

[3]The minus sign is necessary here because, by modern convention, work done by the system on the surroundings is defined as negative. This definition contrasts with the more historical convention that defines the work done by the system on the surroundings as positive.

[4]We could devise a system in which the external pressure does, in fact, change as a function of the volume of the gas. For example, consider a closed system in which a gas on the external pressure side of the piston is compressed, causing the external pressure to increase as the piston moves out. In this case, the functional relationship between the external pressure and the volume of the gas is contrived.

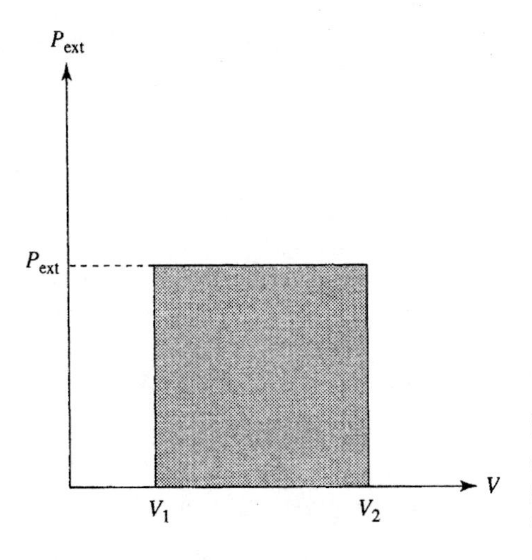

Figure 5-1 Indicator diagram showing *PV* work done by a gas expanding against a constant external pressure.

We can approximate the area under the curve shown in Fig. 5-2 by dividing the area into four rectangles of equal width ΔV. The approximate area under the curve is then just the sum of the areas of the four rectangles shown:

$$A_{\text{approx}} = P_1\,\Delta V + P_2\,\Delta V + P_3\,\Delta V + P_3\,\Delta V$$

$$= \sum_{i=1}^{4} P_i \Delta V \tag{5-4}$$

If we extend this process further—that is, if we divide the area under the curve into more and more rectangles of smaller and smaller width ΔV—the sum approaches a fixed value as N approaches infinity. Without proof, we shall define this

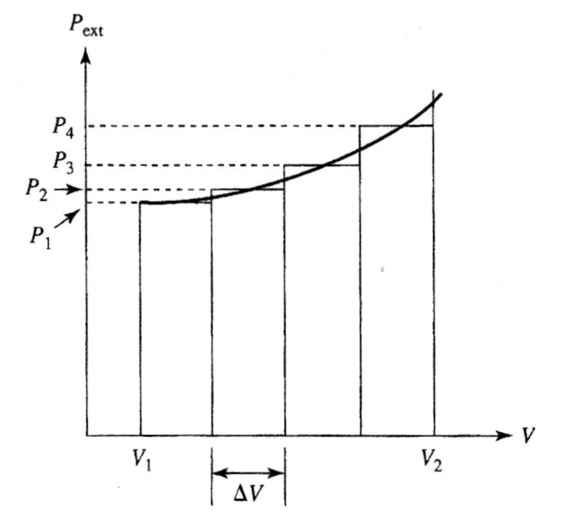

Figure 5-2 Indicator diagram showing *PV* work done by a gas against a changing external pressure.

limiting fixed value of the summation as the true area under the curve in the interval between V_1 and V_2. Hence, we can write

$$A = \lim_{N \to \infty} \sum_{i=1}^{N} P_i \Delta V \tag{5-5}$$

However, because, as N approaches infinity, ΔV approaches zero, we also can write

$$A = \lim_{\Delta V \to 0} \sum_{i=1}^{N} P_i \Delta V \tag{5-6}$$

But, by definition,

$$\lim_{\Delta V \to 0} \sum_{i=1}^{N} P_i \Delta V = \int_{V_1}^{V_2} P \, dV \tag{5-7}$$

where the symbol $\int_{V_1}^{V_2}$ is read "the integral from V_1 to V_2", and V_1 and V_2 are called the *limits of integration*. It follows that

$$A = \int_{V_1}^{V_2} P_{\text{ext}} \, dV \tag{5-8}$$

The integral in Equation (5-8) is called a *definite integral*, because it has a fixed value in the interval between V_1 and V_2. We see, then, that the integral is the summation of an infinite number of infinitesimally small slices, or elements, of area.

Definite integrals are evaluated exactly the same as are indefinite integrals, with the following additional steps: After the integral is found by using the methods outlined in previous sections, the upper limit is substituted for the variable in the integral, and the integral is evaluated. Then the lower limit is substituted for the variable in the integral, and the integral is reevaluated. The value of the definite integral is determined by subtracting the value found by using the lower limit from the value found by using the upper limit. Note that the procedure causes the constant of integration to vanish.

Examples

(a) Evaluate $\int_{3}^{10} x^2 \, dx$.

$$\int_{3}^{10} x^2 \, dx = \left[\frac{x^3}{3} + C\right]_3^{10} = \left[\left(\frac{1000}{3} + C\right) - \left(\frac{27}{3} + C\right)\right]$$

$$= 333 - 9 = 324$$

(b) Evaluate $\int_0^\pi \sin\theta\, d\theta$.

$$\int_0^\pi \sin\theta\, d\theta = [-\cos\theta]_0^\pi = [-\cos\pi - (-\cos 0)] = [1 + 1] = 2$$

Since the constant of integration vanishes, we generally do not include it when we evaluate definite integrals.

(c) Evaluate $\frac{2}{L}\int_0^L \sin^2\left(\frac{n\pi x}{L}\right) x\, dx$.

From the table of integrals in Appendix II,

$$\int x \sin^2(ax + b)\, dx = \frac{x^2}{2} - \frac{x \sin 2(ax + b)}{4a} - \frac{\cos 2(ax + b)}{8a^2} + C;$$

$$a = \frac{n\pi}{L}, b = 0$$

$$\frac{2}{L}\int_0^L \sin^2\frac{n\pi x}{L} x\, dx = \frac{2}{L}\left[\frac{x^2}{4} - \frac{x \sin 2n\pi x/L}{4(n\pi/L)} - \frac{\cos 2n\pi x/L}{8(n\pi/L)^2}\right]_0^L$$

$$= \frac{2}{L}\left[\frac{L^2}{4} - \frac{L \sin 2n\pi}{4(n\pi/L)} - \frac{\cos 2n\pi}{8(n\pi/L)^2}\right.$$

$$\left. -0 + \frac{L \sin 0}{4(n\pi/L)} + \frac{\cos 0}{8(n\pi/L)^2}\right] = \frac{L}{2}$$

5-6 LINE INTEGRALS

Having defined the integral as representing the area under the curve, we now ask whether it is possible to evaluate the integral described in Equation (5-8). Certainly, we can evaluate it numerically, as we did before, by breaking up the integral into little pieces and evaluating the areas of the little pieces. This type of numerical integration is discussed in Chapter 11. But can we evaluate the integral in Equation (5-8), using the analytical methods described in Sections 5-2 and 5-3? For the case of P_{ext} versus V, the answer is no, and the reason is as follows: Integrals of the general type

$$A = \int_{x_1}^{x_2} y\, dx \tag{5-9}$$

are called *line integrals*, because such integrals represent the area under a specific curve (path) connecting x_1 to x_2. Such an integral can be evaluated analytically (i.e., by finding the antiderivative) only if an equation (representing the path) $y = f(x)$ is known, since, under these circumstances, $\int_{x_1}^{x_2} f(x)\, dx$ contains only one variable. If y is not a function of x (as it is not the case of P_{ext} versus V), or if y is a function of x, but the functional dependence of y on x is not known or cannot be integrated, or if y is a function of more than one variable that changes with x, then the line integral cannot be evaluated analytically, and one must resort to a numerical or graphical method of integration in order to evaluate the integral (Chapters 11 and 12).

We sometimes can get around the problem by imposing special conditions on y. For example, the integral in Equation (5-8) can be evaluated analytically if we assume that the external pressure on the gas is constant. Under these circumstances, the constant P_{ext} can be brought out of the integral, and we have

$$A = \int_{V_1}^{V_2} P_{ext}\, dV = P_{ext} \int_{V_1}^{V_2} dV = P_{ext}(V_2 - V_1)$$

which we recognize to be the area of the rectangle shown in Fig. 5-1.

A second way to evaluate the integral in Equation (5-8) is found in the concept of reversibility, which we described in the previous chapter. If the expansion of the gas is reversible, then, for all practical purposes, the gas pressure equals the external pressure at all points in the expansion. This relationship gives

$$A = \int_{V_1}^{V_2} P_{ext}\, dV = \int_{V_1}^{V_2} P(T, V)\, dV$$

Even under these conditions, however, the integral cannot be evaluated, since P is a function of more than one variable. Each value of temperature will describe a specific path leading the system from P_1V_1 to P_2V_2. But if we further stipulate that the temperature of the system is constant, then the integral can be evaluated. If the gas is ideal under isothermal conditions, we can write

$$A = \int_{V_1}^{V_2} P_{gas}\, dV = \int_{V_1}^{V_2} \frac{nRT}{V}\, dV = nRT \int_{V_1}^{V_2} \frac{dV}{V}$$

$$= nRT[\ln V_2 - \ln V_1] = nRT \ln \frac{V_2}{V_1}$$

$$\text{work} = -A = -nRT \ln \frac{V_2}{V_1}$$

Examples

(a) Find the work done when 5.00 moles of a gas obeying van der Waals equation are allowed to expand isothermally and reversibly at 298.2 K from a volume of 1.00 liter to a volume of 5.00 liters. Take $a = 1.358\ \ell^2 \cdot \text{bar mol}^{-1}$ and $b = 0.02789\ \ell \cdot \text{mol}^{-1}$ in van der Waals equation.

Solution

$$\left(P + \frac{n^2a}{V^2}\right)(V - nb) = nRT \quad \text{or} \quad P = \frac{nRT}{(V - nb)} - \frac{n^2a}{V^2}$$

$$w = -\int_{V_1}^{V_2} P_{ext}\, dV = -\int_{V_1}^{V_2} P_{gas}\, dV = -\int_{V_1}^{V_2} \left[\frac{nRT}{(V - nb)} - \frac{n^2a}{V^2}\right] dV$$

$$= -\left[\left(nRT \ln\frac{V_2 - nb}{V_1 - nb}\right) + n^2a\left(\frac{1}{V_2} - \frac{1}{V_1}\right)\right]$$

$$= -nRT \ln\frac{V_2 - nb}{V_1 - nb} - n^2a\left(\frac{1}{V_2} - \frac{1}{V_1}\right)\left(\frac{8.314}{0.08314}\right)$$

(Some of you may be tempted to cancel out the nb's in the logarithmic term in the preceding equation. Don't!) The ratio 8.314/0.08314 in the second term is required to change the units of the second term from liter · bars to joules. We have

$$w = -(5.00)(8.314)(298.2) \ln\frac{5.00 - (5.00)(0.02789)}{1.00 - (5.00)(0.02789)}$$

$$-(5.00)^2(1.358)\left(\frac{1}{5.00} - \frac{1}{1.00}\right)(100.0)$$

$$= -21462 + 2716 = -18746 \text{ J}$$

(b) The change in enthalpy as a function of temperature is given by the equation

$$\Delta H = \int_{T_1}^{T_2} C_P\, dT \quad \text{(per mole)}$$

where C_P is the heat capacity at constant pressure. Find the change in enthalpy for 1 mole of a gas, when the temperature of the gas changes from, say, 298.2 K to 500.0 K.

Solution. We first recognize that the foregoing integral is a line integral. This integral cannot be evaluated unless C_P as a function of T is known. We could assume that C_P is constant and evaluate the integral under that assumption. However, over a large temperature range, it would be a poor assumption. Another way would be to determine the integral numerically, using methods described in Chapter 11. A common analytical approach is to express C_P as a power series in temperature. (Power series are covered in Chapter 6.) While this is not the exact functional relationship between C_P and T, we find that such an approach gives good results:

$$C_P = a + bT + cT^2$$

The constants a, b, and c are known for many common gases. Substituting the preceding equation into that for ΔH yields

$$\Delta H = \int_{T_1}^{T_2} (a + bT + cT^2)\, dT$$

which now can be integrated to give

$$\Delta H = a(T_2 - T_1) + \frac{b}{2}(T_2^2 - T_1^2) + \frac{c}{3}(T_2^3 - T_1^3)$$

We see, then, that the change in enthalpy of the system can be determined by summing over the entire temperature range, one infinitesimal contribution $C_P\, dT$ at a time.

5-7 DOUBLE AND TRIPLE INTEGRALS

In Chapter 4, we saw that functions could be differentiated more than once. Let us consider the inverse of this process: the determination of multiple integrals. The volume of a cylinder is a function of both the radius and the height of the cylinder. That is, $V = f(r, h)$. Let us suppose that we allow the height of the cylinder, h, to change while holding the radius r constant. The integral from $h = 0$ to $h = H$ could then be expressed as

$$\int_0^H f(r, h)\, dh \tag{5-10}$$

But the value of this integral depends on the value of the radius r, and hence the integral could be considered to be a function of r. That is,

$$g(r) = \int_0^H f(r, h)\, dh \tag{5-11}$$

If we now let r vary from $r = 0$ to $r = R$ and integrate over the change, we can write

$$\int_0^R g(r)\, dr = \int_0^R \int_0^H f(r, h)\, dh\, dr \tag{5-12}$$

which is read, "the double integral of $f(r, h)$ from $h = 0$ to $h = H$ and $r = 0$ to $r = R$."

To evaluate the double integral, we integrate $\int_0^H f(r, h)\, dh$ first while holding r constant, which gives us $g(r)$. Then we integrate $\int_0^R g(r)\, dr$ while holding h constant. Such a process is known as *successive partial integration*. For example, let us evaluate $\int_0^R \int_0^H 2\pi r\, dh\, dr$. First, we get

$$g(r) = \int_0^H 2\pi r\, dh = 2\pi r H$$

Next, we integrate

$$\int_0^R g(r)\,dr = \int_0^R 2\pi r H\,dr = \pi R^2 H$$

which we recognize as the volume of a cylinder with $r = R$ and $h = H$.

The preceding argument can be extended to the triple integral—or any multiple integral, for that matter. For example, let us evaluate the triple integral describing the volume element in spherical polar coordinates (see Chapter 1):

$$\int_0^{2\pi}\int_0^{\pi}\int_0^R r^2 \sin\theta\,dr\,d\theta\,d\phi$$

Note that the limits of ϕ are from 0 to 2π, while the limits of θ are only from 0 to π. If both integrals were evaluated from 0 to 2π, we would be counting some angles twice. Evaluating the ϕ integral first, we have

$$\int_0^{2\pi} r^2 \sin\theta\,d\phi = r^2 \sin\theta[\phi]_0^{2\pi} = 2\pi r^2 \sin\theta$$

$$\int_0^{\pi} 2\pi r^2 \sin\theta\,d\theta = 2\pi r^2[-\cos\theta]_0^{\pi} = 2\pi r^2[-(-1) + 1] = 4\pi r^2$$

which we recognize as the surface area of a sphere. Finally,

$$\int_0^R 4\pi r^2\,dr = \frac{4}{3}\pi R^3$$

which we recognize as the volume of a sphere.

SUGGESTED READINGS

1. BRADLEY, GERALD L., and SMITH, KARL J. *Calculus*. Upper Saddle River, NJ: Prentice Hall, 1995.
2. VARBERG, DALE, and PURCELL, EDWIN J. *Calculus*, 7th ed. Upper Saddle River, NJ: Prentice Hall, 1997.

PROBLEMS

1. Evaluate the following integrals (consider all uppercase letters to be constants):

(a) $\int 5x^3\,dx$

(b) $\int \frac{1}{x^3}dx$

(f) $\int P\,dv$

(g) $\int \frac{RT}{p}dp$

(c) $\int \sin 2x \, dx$

(d) $\int (3x + 5)^2 4x \, dx$

(e) $\int 4e^{2x} \, dx$

(h) $\int Mv \, dv$

(i) $\int \frac{Q^2}{r^2} dr$

(j) $\int \cos(2\pi W t) \, dt$

2. Evaluate the following integrals, using the table of integrals found in Appendix II as needed (consider all uppercase letters to be constants):

(a) $\int e^{-4x} \, dx$

(b) $\int (x^2 - A^2) \, dx$

(c) $\int \sqrt{(x^2 - A^2)} \, dx$

(d) $\int (x^4 - 2x^2 + 4)x^3 \, dx$

(e) $\int \sin^2\left(\frac{N\pi x}{A}\right) dx$

(f) $\int \sin^2\left(\frac{N\pi x}{A}\right) x \, dx$

(g) $\int \left(\frac{-\Delta H}{Rt^2}\right) dt$

(h) $\int e^x \sin x \, dx$

(i) $\int \sin^2(2\pi W t) \, dt$

(j) $\int \cos^3 \phi \sin \phi \, d\phi$

(k) $\int \cos^5 \theta \, d\theta$

(l) $\int \sin^4(3x + 4) \, dx$

(m) $\int x^3 \cos 2x \, dx$

(n) $\int \frac{dx}{(4 - x)(3 - x)}$

(o) $\int \left(\frac{\Delta H}{t^2} + \frac{A}{t} + \frac{B}{2} + \frac{C}{3} t\right) dt$

(p) $\int \frac{C_P}{t} dt$

(q) $\int \frac{dx}{(A - x)^n}$

(r) $\int re^{-ar} \, dr$

(s) $\int e^{-\varepsilon/KT} \, d\varepsilon$

(t) $\int \frac{d(a)}{(a)} = -\int K \, dt$

3. Evaluate the following definite integrals, using the table of indefinite and definite integrals found in Appendix II as needed:

(a) $\int_{T_1}^{T_2} \left(a + bT + cT^2 + \frac{d}{T}\right) dT$; a, b, c, and d constants

(b) $\int_{P_1}^{P_2} \frac{RT}{P} dP$; R and T constants

(c) $\int_0^{2\pi} d\phi$

(d) $\int_{T_1}^{T_2} \frac{\Delta H}{RT^2} dT$; ΔH and R constants

(e) $\int_{V_1}^{V_2}\left(\frac{nRT}{V-nb}-\frac{n^2a}{V^2}\right)dV$; a, b, n, R, and T constants

(f) $\int_0^{\pi/2}\sin^2\theta\cos\theta\,d\theta$

(g) $\int_0^a x^2\sin^2\left(\frac{n\pi x}{a}\right)dx$; n, π, and a constants

(h) $\int_0^\infty x^2e^{-ax^2}\,dx$; a constant (see table of definite integrals)

(i) $\int_0^\infty e^{-2r/a_0}r\,dr$; a_0 constant (see table of definite integrals)

(j) $\int_0^\infty e^{-mv^2/2kT}v^3\,dv$; m, k, and T constants (see table of definite integrals)

(k) $\int_0^\infty (2J+1)e^{-a(J^2+J)}\,dJ$; a constant (see table of definite integrals)

4. Consider the ideal gas law $P = nRT/V$, where, in this case, n, R, and T are constants. Prepare a graph of P versus V, choosing suitable coordinates (for $n = 1.00$ mole, $R = 0.08314\ \ell\cdot\text{bar mol}^{-1}\cdot\text{K}^{-1}$, and $T = 298\text{K}$), from a volume of $V = 1.00$ liter to a volume of $V = 10.00$ liters. Now consider the area under the curve from a volume of $V = 2.00$ liters to a volume of $V = 6.00$ liters. Determine the approximate area graphically by breaking it up into four rectangles of equal width ΔV; compare your answer with that found by analytically integrating the function between these limits of integration.

5. Evaluate the following multiple integrals, using the table of integrals as needed:

(a) $\iint yx^2\,dx\,dy$

(b) $\iint (x^2+y^2)\,dx\,dy$

(c) $\iint y\ln x\,dx\,dy$

(d) $\iiint x^2\ln ye^{2x}\,dx\,dy\,dz$

(e) $\int_0^{\pi/2}\int_0^2 r\cos\theta\,dr\,d\theta$

(f) $\int_0^{2\pi}\int_0^{\pi}\int_0^{V} v^2\sin\theta\,dv\,d\theta\,d\phi$

(g) $\int_0^\infty\int_0^\infty\int_0^\infty e^{\frac{h^2}{8mkT}\left(\frac{n_x^2}{a^2}+\frac{n_y^2}{b^2}+\frac{n_z^2}{c^2}\right)}\,dn_x\,dn_y\,dn_z$; a, b, c, m, k, h, and T constants (see table of definite integrals)

6. The equation of a straight line passing through the origin of a Cartesian coordinate system is $y = mx$, where m is the slope of the line. Show that the area of a triangle made up of this line and the x-axis between $x = 0$ and $x = a$ is $A = \frac{1}{2}ay$.

7. The Kirchhoff equation for a chemical reaction relating the variation: ΔH of a reaction with the absolute temperature is

$$\left[\frac{\partial(\Delta H)}{\partial T}\right]_P = \Delta C_P$$

where ΔC_P is the change in the heat capacity at constant pressure for the reaction. Expressing ΔC_P as a truncated power series in T, namely,

$$\Delta C_p = a + bT + cT^2$$

derive an equation for ΔH as a function of temperature. (*Hint*: Write the derivative in differential form, and do not integrate between definite limits. Do not forget the constant of integration!)

8. The Gibbs–Helmholtz equation for a chemical reaction is

$$\left[\frac{\partial(\Delta G/T)}{\partial T}\right]_P = -\frac{\Delta H}{T^2}$$

where ΔG is the change in the Gibbs free energy for the reaction, ΔH is the change in enthalpy attending the reaction, and T is the absolute temperature. Using the expansion for ΔH found in Problem 7, derive an expression for ΔG as a function of temperature. (*Hint*: Write the derivative in differential form. Do not integrate between definite limits. Do not forget the constant of integration.)

9. Determine the probability of finding a particle confined to a field-free one-dimensional box in the state $n = 1$ at $x = L/2$ in the range $L/2 \pm 0.05L$, where L is the width of the box. Use the formula

$$\text{Probability} = \frac{2}{L}\int_{L/2-0.05L}^{L/2+0.05L} \sin^2\frac{\pi x}{L}\,dx$$

10. Determine the probability of finding an electron in the $1s$ state of the hydrogen atom at $r = a_0$ in the range $a_0 \pm 0.005a_0$, where a_0 is the Bohr radius. Use the formula

$$\text{Probability} = 4\left(\frac{1}{a_0}\right)^3 \int_{a_0-0.005a_0}^{a_0+0.005a_0} e^{-2r/a_0} r^2\,dr$$

11. Find the expectation value

$$\langle x \rangle = 4\left(\frac{1}{a_0}\right)^3 \int_0^\infty e^{-2r/a_0}\, r^3\, dr$$

for an electron in the $1s$ state of the hydrogen atom.

12. The differential volume element in cylindrical coordinates is $dV = r\, d\theta\, dr\, dz$. Show that if r goes from 0 to R, θ from 0 to 2π, and z from 0 to h, the volume of a cylinder is $V = \pi R^2 h$.

Infinite Series

6-1 INTRODUCTION

We saw in preceding chapters that it is sometimes useful to express a function as a sum of terms called a *series*—for example,

$$e^x = 1 + x + \frac{x^2}{2} + \frac{x^3}{6} + \frac{x^4}{24} + \ldots$$

or

$$\sin x = x - \frac{x^3}{6} + \frac{x^5}{120} - \frac{x^7}{5040} + - \ldots$$

The three dots (called an *ellipsis*) at the end of each of these series signify that the number of terms in the series is endless. Therefore, such a series is called an *infinite series*, expressed in general form as

$$\sum_{n=1}^{\infty} a_n = a_1 + a_2 + a_3 + \ldots + a_n + \ldots \qquad (6\text{-}1)$$

Suppose we have a sequence of terms $u_1, u_2, u_3, \ldots$ and that we let S_n be the sum of the first n terms. That is,

$$S_n = u_1 + u_2 + u_3 + \ldots + u_n \qquad (6\text{-}2)$$

The term S_n is called the *nth partial sum*. Thus, an infinite series is an infinite sequence of partial sums $S_1, S_2, S_3, \ldots$. If, as n approaches infinity, the sum S_n approaches a definite, finite value S, then the series *converges* (or is *convergent*), and we say that

$$S = \lim_{n \to \infty} S_n \qquad (6\text{-}3)$$

If, however, as n approaches infinity, S_n does not have a definite, finite value, but increases without limit, then the series *diverges* (or is *divergent*). Examples of convergent series and their sums are

$$S = 1 - \frac{1}{2} + \frac{1}{3} - \frac{1}{4} + - \ldots = \sum_{x=1}^{\infty} \frac{(-1)^{x-1}}{x} = \ln 2$$

$$S = 1 + a + a^2 + a^3 + \ldots = \sum_{x=1}^{\infty} a^x = \frac{1}{1-a}; \quad |a| < 1$$

and

$$S = a + 2a^2 + 3a^3 + \ldots = \sum_{x=1}^{\infty} x a^x = \frac{a}{(1-a)^2}; \quad |a| < 1$$

6-2 TESTS FOR CONVERGENCE AND DIVERGENCE

A number of methods are used to determine whether an infinite series converges or diverges. One common method is to examine the nth partial sum and find its limit as $n \to \infty$. For example, consider the series

$$S = \sum_{x=1}^{\infty} \frac{1}{x(x+1)} = \frac{1}{2} + \frac{1}{6} + \frac{1}{12} + \frac{1}{20} + \ldots$$

The first four partial sums are

$$S_1 = \frac{1}{2}, \quad S_2 = \frac{2}{3}, \quad S_3 = \frac{3}{4}, \quad S_4 = \frac{4}{5}$$

It is easy to verify that the sum for any n is given by the equation

$$S_n = \frac{n}{n+1}$$

Taking the limit as $n \to \infty$, we have

$$S = \lim_{n \to \infty} \left(\frac{n}{n+1} \right) = 1$$

The series converges with a sum equal to unity.

Examples

(a) Determine whether the series

$$\sum_{n=1}^{\infty} \frac{1}{3^n} = \frac{1}{3} + \frac{1}{9} + \frac{1}{27} + \frac{1}{81} + \ldots$$

converges or diverges.

Solution. The first four partial sums are

$$S_1 = \frac{1}{3}, \quad S_2 = \frac{4}{9}, \quad S_3 = \frac{13}{27}, \quad S_4 = \frac{40}{81}$$

We can see that in this case

$$S_n = \frac{1}{2}\left(\frac{3^n - 1}{3^n}\right)$$

Therefore,

$$\lim_{n \to \infty} \frac{1}{2}\left(\frac{3^n - 1}{3^n}\right) = \frac{1}{2}$$

The series converges.

(b) Determine whether the series

$$\sum_{n=1}^{\infty} \frac{1}{2n} = \frac{1}{2} + \frac{1}{4} + \frac{1}{6} + \frac{1}{8} + \ldots$$

converges or diverges.

Solution. The first four partial sums are

$$S_1 = \frac{1}{2}, \quad S_2 = \frac{3}{4}, \quad S_3 = \frac{11}{12}, \quad S_4 = \frac{25}{24}$$

The equation for determining the partial sums is not obvious here. We sometimes can determine it by integration:

$$\int_1^{\infty} \frac{dn}{2n} = \lim_{b \to \infty} \int_1^b \frac{dn}{2n} = \lim_{b \to \infty} \left[\frac{1}{2} \ln 2b - \frac{1}{2} \ln 2\right] = \infty$$

The sum diverges as $\frac{1}{2} \ln 2n$, so the series diverges.

Geometric Series A series $\sum_{n=0}^{\infty} a_n$ is said to be a geometric series if every term after the first term is a fixed multiple of the term before it, or, mathematically,

$$\sum_{n=0}^{\infty} a_n = a_0 + a_0 u + a_0 u^2 + a_0 u^3 + \ldots = \sum_{n=0}^{\infty} u^n a_0 \tag{6-4}$$

The sum of a geometric series is given without proof as

$$S = \frac{a_0}{1 - u} \tag{6-5}$$

If $|u| < 1$, the series converges to S. If $|u| > 1$, then S does not exist and the series diverges.

Example

Determine whether the series

$$\sum_{n=0}^{\infty} \frac{1}{2^n} = 1 + \frac{1}{2} + \frac{1}{4} + \frac{1}{8} + \frac{1}{16} + \ldots$$

converges or diverges. If the series converges, what is the sum?

Solution. We recognize this series as a geometric series with $a_0 = 1$ and $u = 1/2$. Since $u < 1$, the series converges:

$$S = \frac{1}{1 - \frac{1}{2}} = 2$$

Comparison Test A common way to determine whether a series is convergent or divergent is to compare it with series that are known either to converge or to diverge. Consider the following series:

(1) The geometric series

$$a + ar + ar^2 + ar^3 + \ldots + ar^n + \ldots$$

converges when $r < 1$ and diverges when $r \geq 1$.

(2) The series

$$1 + \frac{1}{2^p} + \frac{1}{3^p} + \frac{1}{4^p} + \ldots + \frac{1}{n^p} + \ldots$$

converges when $p > 1$.

(3) The harmonic series

$$1 + \frac{1}{2} + \frac{1}{3} + \frac{1}{4} + \ldots + \frac{1}{n} + \ldots$$

always diverges.

(4) The series

$$\frac{1}{a(a+1)} + \frac{1}{(a+1)(a+2)} + \ldots + \frac{1}{(a+n-1)(a+n)} + \ldots$$

where $a > 0$, always converges.

Example

Determine whether the series

$$\sum_{n=1}^{\infty} \frac{n+1}{n} = \frac{2}{1} + \frac{3}{2} + \frac{4}{3} + \frac{5}{4} + \ldots$$

converges or diverges.
We compare this series with the harmonic series $1 + \frac{1}{2} + \frac{1}{3} + \frac{1}{4} + \ldots$. Note that each term in the given series is always greater than the corresponding term in the harmonic series. Therefore, the given series diverges.

Ratio Test Another method used to determine whether a series is convergent or divergent is the so-called *ratio test.* Consider the ratio of the $(n+1)$st term to the nth term in the infinite series $u_1 + u_2 + u_3 + \ldots$. We find that if

$$\lim_{n\to\infty} \left| \frac{u_{n+1}}{u_n} \right| < 1$$

the series converges. If, however,

$$\lim_{n\to\infty} \left| \frac{u_{n+1}}{u_n} \right| > 1$$

the series diverges. Note that when

$$\lim_{n\to\infty}\left|\frac{u_{n+1}}{u_n}\right| = 1$$

the test fails and another method must be used.

Example

Determine whether the series

$$\frac{1}{2} + \frac{2}{2^2} + \frac{3}{2^3} + \frac{4}{2^4} + \ \ldots$$

converges or diverges.

Solution. In this series, the general term is $u_n = n/2^n$. Hence,

$$u_{n+1} = \frac{n+1}{2^{n+1}}$$

and

$$\lim_{n\to\infty}\left|\frac{u_{n+1}}{u_n}\right| = \lim_{n\to\infty}\left|\frac{n+1}{2^{n+1}}\cdot\frac{2^n}{n}\right| = \lim_{n\to\infty}\left|\frac{n+1}{2n}\right|$$

However, since

$$\lim_{n\to\infty}\left|\frac{n+1}{2n}\right| = \frac{1}{2}$$

and this is less than unity, the series must converge.

6-3 POWER SERIES REVISITED

One of the most useful mathematical tools in applied mathematics is the *power series*, an infinite series having the form

$$a_0 + a_1x + a_2x^2 + a_3x^3 + \ \ldots \ + a_nx^n + \ \ldots \ = \sum_{n=0}^{\infty} a_nx^n \qquad (6\text{-}6)$$

Power series, like other infinite series, can be either convergent or divergent; however, whether a series converges or diverges depends on the value of x.

We can determine the value of x for which a series converges or diverges by applying the ratio test. We find that the series converges for all values of x in the interval

$$|x| < \lim_{n\to\infty} \left| \frac{a_{n-1}}{a_n} \right|$$

and that the series diverges for all other values of x—that is, for

$$|x| > \lim_{n\to\infty} \left| \frac{a_{n-1}}{a_n} \right|$$

At the endpoints of the interval, that is, at

$$|x| = \lim_{n\to\infty} \left| \frac{a_{n-1}}{a_n} \right|$$

the test fails.

Example

Determine for which values of x the series

$$\sum_{n=0}^{\infty} a_n x^{n+1} = a_0 x + a_1 x^2 + a_2 x^3 + \ldots = x + 2x^2 + 3x^3 + \ldots$$

converges.

Solution. In this case, $a_0 = 1$, $a_1 = 2$, $a_2 = 3$, and so on. Hence,

$$a_n = n + 1 \quad \text{and} \quad a_{n-1} = n$$

Thus,

$$\lim_{n\to\infty} \left| \frac{a_{n-1}}{a_n} \right| = \lim_{n\to\infty} \left| \frac{n}{n+1} \right| = 1$$

The series converges for $|x| < 1$, that is, in the interval $-1 < x < +1$. Likewise, the series diverges for $|x| > 1$, that is, for $x < -1$ or $x > +1$. When $x = 1$, the series becomes

$$1 + 2 + 3 + 4 + 5 + \ldots$$

which diverges. When $x = -1$, the series becomes

$$-1 + 2 - 3 + 4 - 5 + - \ldots$$

which also diverges. Hence, the interval of convergence does not include the endpoints.

6-4 MACLAURIN AND TAYLOR SERIES

In this chapter, we consider several methods of expanding functions in infinite series. Two that are particularly useful in physical chemistry are the power series known as the *Maclaurin series* and the *Taylor series*. Let us consider the Maclaurin series first.

Suppose that a function $f(x)$ can be expanded in a power series

$$f(x) = a_0 + a_1x + a_2x^2 + a_3x^3 + \dots \tag{6-7}$$

Suppose further that the function has continuous derivatives of all orders. Let us evaluate $f(x)$ and its derivatives at $x = 0$:

$$\begin{aligned}
f(x) &= a_0 + a_1x + a_2x^2 + a_3x^3 + \dots; f(0) = a_0 \\
f'(x) &= a_1 + 2(1)a_2x + (3)(1)a_3x^2 + \dots; f'(0) = a_1 \\
f''(x) &= (2)(1)a_2 + (3)(2)(1)a_3x + \dots; f''(0) = 2!a_2 \\
f^n(x) &= n!a_n + (n+1)!a_{n+1}x + \dots; f^n(0) = n!a_n
\end{aligned}$$

Substituting these values for the coefficients back into Equation (6-7) gives

$$f(x) = f(0) + f'(0)x + \frac{f''(0)}{2!}x^2 + \dots + \frac{f^n(0)}{n!}x^n + \dots \tag{6-8}$$

which is known as the Maclaurin series. To illustrate the use of this series, let us expand the function $f(x) = \sin x$ in a Maclaurin series:

$$\begin{aligned}
f(x) &= \sin x & f(0) &= 0 \\
f'(x) &= \cos x & f'(0) &= 1 \\
f''(x) &= -\sin x & f''(0) &= 0 \\
&\vdots & &\vdots
\end{aligned}$$

$$\sin x = x - \frac{x^3}{3!} + \frac{x^5}{5!} - \frac{x^7}{7!} + - \dots$$

Note that the sine function involves the odd powers of x. For this reason and others, the sine function is referred to as an *odd* function. An important property of an odd function is that $f(-x) = -f(x)$. A similar expansion of $\cos x$ will show that the cosine function involves the even powers of x. Thus, the cosine function is referred to as an *even* function. An important property of an even function is that $f(-x) = f(x)$.

PROBLEM. Expand the function $\ln(1 + x)$ in a Maclaurin series. We have

$$f(x) = \ln(1 + x) \qquad f(0) = 0$$

$$f'(x) = \frac{1}{1 + x} \qquad f'(0) = 1$$

$$f''(x) = \frac{-1}{(1 + x)^2} \qquad f(0) = -1$$

$$\vdots \qquad \vdots$$

$$\ln(1 + x) = x - \frac{x^2}{2} + \frac{x^3}{3} - + \ldots$$

Let us consider, now, a function that can be expanded in the series

$$f(x) = c_0 + c_1(x - a) + c_2(x - a)^2 + c_3(x - a)^3 + \ldots \tag{6-9}$$

where a is some constant. Assume, as we did before, that the function $f(x)$ has continuous derivatives of all orders. Let us now evaluate the function and its derivatives at $x = a$, rather than at $x = 0$:

$$f(x) = c_0 + c_1(x - a) + c_2(x - a)^2 + c_3(x - a)^3 + \ldots; \quad f(a) = c_0$$
$$f'(x) = c_1 + 2c_2(x - a) + 3c_3(x - a)^2 + \ldots; \quad f'(a) = c_1$$
$$f''(x) = 2c_2 + 3(2)(1)c_1(x - a) + \ldots; \quad f''(a) = 2!c_0$$
$$f^n(x) = n!c_n + (n + 1)!c_{n+1}(x - a) + \ldots; \quad f^n(a) = n!c_n$$

Substituting these values for the coefficients back into Equation (6-9) gives

$$f(x) = f(a) + f'(a)(x - a) + \frac{f''(a)}{2!}(x - a)^2 + \ldots + \frac{f^n(a)}{n!}(x - a)^n + \ldots \tag{6-10}$$

which is known as a Taylor's series.

PROBLEM. Expand the function $f(x) = e^x$ in powers of $(x + 2)$. Since $f(x) = e^x$ and $(x + 2) = (x - a)$, or $a = -2$, we have

$$f(x) = e^x \qquad f(a) = e^{-2}$$

$$f'(x) = e^x \qquad f'(a) = e^{-2}$$

$$f''(x) = e^x \qquad f''(a) = e^{-2}$$

$$\vdots \qquad \vdots$$

$$f(x) = e^{-2}\left[1 + (x + 2) + \frac{1}{2}(x + 2)^2 + \frac{1}{6}(x + 2)^3 + \ldots\right]$$

6-5 FOURIER SERIES AND FOURIER TRANSFORMS

One of the most common mathematical tools in chemistry and physics is the Fourier transform, essentially a mathematical manipulation that reorganizes information. Fourier transforms arise naturally in a number of physical problems. For example, a lens is a Fourier transformer. The human eye and ear act as Fourier transformers by analyzing complex electromagnetic and sound waves, respectively.

A Fourier transform is performed with the use of a Fourier series, expressed in its most general form as the sum of sine and cosine functions:

$$f(x) = \sum_n a_n \sin nx + \frac{1}{2}b_0 + \sum_n b_n \cos nx \tag{6-11}$$

In this equation, the coefficients

$$a_n = \frac{1}{\pi}\int_{-\pi}^{\pi} f(x) \sin nx \, dx \quad \text{and} \quad b_n = \frac{1}{\pi}\int_{-\pi}^{\pi} f(x) \cos nx \, dx \tag{6-12}$$

The coefficients are found from the *orthonormal* behavior of sine and cosine functions, which we will discuss later in the chapter. Fourier series differ from power series expansions in at least two important ways. First, the interval of convergence of a power series is different for different functions; the Fourier series, by contrast, always converges between $-\pi$ and $+\pi$. Second, many functions cannot be expanded in a power series, whereas it is rare to find a function that cannot be expanded in a Fourier series.

It is sometimes useful to change the range of the Fourier series from $(-\pi, \pi)$ to $(-L, L)$. This can be accomplished by replacing the variable x in Equation (6-11) with $\pi x/L$, giving

$$f(x) = \sum_n a_n \sin\frac{n\pi x}{L} + \frac{1}{2}b_0 + \sum_n b_n \cos\frac{n\pi x}{L} \tag{6-13}$$

where now

$$a_n = \frac{1}{L}\int_{-L}^{+L} f(x) \sin\frac{n\pi x}{L}dx \quad \text{and} \quad b_n = \frac{1}{L}\int_{-L}^{+L} f(x) \cos\frac{n\pi x}{L}dx \tag{6-14}$$

We saw in Chapter 1 that sine and cosine functions also can be represented as complex exponential functions. It is easy to show, therefore, that another form of the Fourier series is

$$f(x) = \sum_{-\infty}^{+\infty} c_n e^{in\pi x/L} \tag{6-15}$$

where $c_n = \frac{1}{2L}\int_{-L}^{+L} f(x)e^{-in\pi x/L}\,dx$. Note that, in this case, the coefficients are complex and the series will represent only functions that are periodic. It is possible to modify Equation (6-15) to represent functions that are not periodic. To do this, let $k = n\pi/L$. Now, let us see what happens when we allow L to go to infinity. As L gets larger and larger, k changes in smaller and smaller increments with each change in n. That is, $\Delta k = (\pi/L)\Delta n$. In the limit as Δk goes to zero, k becomes a continuous variable, and the coefficients c_k can be described as continuous functions of k, $c(k)$. Therefore, substituting for $\Delta n = L/\pi\Delta k$ yields

$$f(x) = \lim_{L\to\infty} \sum_{-\infty}^{+\infty} c_n e^{in\pi x/L}\Delta n$$

$$f(x) = \lim_{\Delta k\to 0} \sum_{-\infty}^{+\infty} \frac{Lc(k)}{\pi} e^{ikx}\Delta k$$

But this is just the definition of an integral. Letting $Lc(k)/\pi = g(k)/\sqrt{2\pi}$, we have

$$f(x) = \frac{1}{\sqrt{2\pi}}\int_{-\infty}^{+\infty} g(k)e^{ikx}\,dk \tag{6-16}$$

and

$$g(k) = \frac{1}{\sqrt{2\pi}}\int_{-\infty}^{+\infty} f(x)e^{-ikx}\,dx \tag{6-17}$$

Equation (6-16) is called the *Fourier integral*, and $f(x)$ and $g(k)$ are called *continuous Fourier transforms* of one another. Normally, Fourier transforms are calculated by computer. However, digital computers do not handle continuous data very well. For that reason, the numerical calculation of Fourier transforms by computer requires discrete samples of the function. Therefore, in Chapter 11, we will introduce the *discrete Fourier transform* and illustrate how to calculate a Fourier transform with the use of computer methods.

PROBLEM. Find the Fourier transform of the function

$$f(t) = \begin{cases} 0; & t < -\tau/2 \\ \sqrt{\frac{\pi}{2}}\cos\omega_0 t; & -\tau/2 < t < +\tau/2 \\ 0; & t > +\tau/2 \end{cases}$$

where the cosine wave exists only in the period τ.

The Fourier transform of this function is

$$g(\omega) = \frac{1}{\sqrt{2\pi}} \int_{-\tau/2}^{+\tau/2} e^{-i\omega t} \sqrt{\frac{\pi}{2}} \cos \omega_0 t \, dt$$

Note that, since the function equals zero outside the range $(-\tau/2, \tau/2)$, we need only integrate from $-\tau/2$ to $+\tau/2$, rather than from $-\infty$ to $+\infty$. Using Euler's relation (see Chapter 1), we have

$$\begin{aligned} g(\omega) &= \frac{1}{2} \int_{-\tau/2}^{+\tau/2} (\cos \omega t - i \sin \omega t) \cos \omega_0 t \, dt \\ &= \frac{1}{2} \int_{-\tau/2}^{+\tau/2} \cos \omega t \cos \omega_0 t \, dt - \frac{i}{2} \int_{-\tau/2}^{+\tau/2} \sin \omega t \cos \omega_0 t \, dt \end{aligned}$$

These integrals can be found in the table of integrals in Appendix II. Integration yields

$$\begin{aligned} g(\omega) &= \frac{1}{2} \left[\frac{\sin(\omega - \omega_0)t}{2(\omega - \omega_0)} + \frac{\sin(\omega + \omega_0)t}{2(\omega + \omega_0)} \right. \\ &\quad \left. + \frac{i \cos(\omega - \omega_0)t}{2(\omega - \omega_0)} + \frac{i \cos(\omega + \omega_0)t}{2(\omega + \omega_0)} \right]_{-\tau/2}^{+\tau/2} \\ &= \frac{1}{2} \left[\frac{\sin[(\omega - \omega_0)(\tau/2)]}{2(\omega - \omega_0)} + \frac{\sin[(\omega + \omega_0)(\tau/2)]}{2(\omega + \omega_0)} \right. \\ &\quad + \frac{i \cos[(\omega - \omega_0)(\tau/2)]}{2(\omega - \omega_0)} + \frac{i \cos[(\omega + \omega_0)(\tau/2)]}{2(\omega + \omega_0)} \\ &\quad - \frac{\sin[(\omega - \omega_0)(-\tau/2)]}{2(\omega - \omega_0)} - \frac{\sin[(\omega + \omega_0)(-\tau/2)]}{2(\omega + \omega_0)} \\ &\quad \left. - \frac{i \cos[(\omega - \omega_0)(-\tau/2)]}{2(\omega - \omega_0)} - \frac{i \cos[(\omega + \omega_0)(-\tau/2)]}{2(\omega + \omega_0)} \right] \end{aligned}$$

or

$$g(\omega) = \frac{\sin[(\omega - \omega_0)(\tau/2)]}{2(\omega - \omega_0)} + \frac{\sin[(\omega + \omega_0)(\tau/2)]}{2(\omega + \omega_0)}$$

since $\sin(-x) = -\sin x$ and $\cos(-x) = \cos x$. Note that the Fourier transform transformed the function from a time domain t to a frequency domain ω.

Simpler forms of the Fourier series are possible. For example, if the function to be expanded is an *odd* function about $x = 0$ [i.e., $f(x) = -f(-x)$], then the b coefficients in Equation (6-11) vanish and the Fourier series becomes

$$f(x) = \sum_{n=1}^{\infty} a_n \sin nx \tag{6-18}$$

where

$$a_n = \frac{2}{\pi}\int_0^{\pi} f(x) \sin nx \, dx \tag{6-19}$$

Likewise, if the function $f(x)$ is an *even* function about $x = 0$ [i.e., $f(x) = f(-x)$], then the a coefficients in Equation (6-11) vanish and the Fourier series becomes

$$f(x) = \frac{1}{2}b_0 + \sum_{n=1}^{\infty} b_n \cos nx \tag{6-20}$$

where

$$b_n = \frac{2}{\pi}\int_0^{\pi} f(x) \cos nx \, dx \tag{6-21}$$

Of all the wave functions that can be used to describe a quantum mechanical system, the most important are those wave functions having a definite total energy. There are at least two major reasons for this. The first, obviously, is because such functions indeed represent states with definite total energy. A more subtle reason, however, is that other wave functions can always be defined as linear combinations of definite-energy wave functions. For example, the wave-mechanical solutions of equations describing a particle confined to a one-dimensional box between $x = 0$ and $x = L$ are

$$\psi = \sqrt{\frac{2}{L}} \sin\frac{n\pi x}{L} \tag{6-22}$$

where n is a positive integer (not including zero). This complete set of functions can be used as a natural basis set for a Fourier expansion. The wave functions described in Equation (6-22) have at least two important properties. First, the solutions are said to be *normalized*. The condition for normalization is

$$\int_{\text{all space}} \psi_i{}^*\psi_i \, d\tau = 1 \tag{6-23}$$

where ψ^* denotes the complex conjugate of ψ and $d\tau$ is the differential volume element for the coordinate system. Second, any two different solutions are said to be *orthogonal*. The condition for orthogonality is

$$\int_{\text{all space}} \psi_i^* \psi_j \, d\tau = 0 \tag{6-24}$$

Taken together, the preceding two conditions are said to form a *complete orthonormal* set, sometimes designated as

$$\int_{\text{all space}} \psi_i^* \psi_j \, d\tau = \delta_{ij}; \quad \delta_{ij} = \begin{cases} 1, & i = j \\ 0, & i \neq j \end{cases} \tag{6-25}$$

The term δ_{ij} is called *Kronecker delta*.

We can use this complete set given in Equation (6-22) as a basis set for a Fourier series expansion of any function in the interval from 0 to L. For example, let us expand the function $f(x) = 3x$ in a Fourier series, using the solutions to the particle-in-the-box problem as a basis set:

$$f(x) = \sum_n a_n \psi_n = \sum_n a_n \sqrt{\frac{2}{L}} \sin \frac{n\pi x}{L} \tag{6-26}$$

We shall see in Chapter 7 that Equation (6-26) is equivalent to taking a linear combination of the wave functions.

To find the coefficients, we make use of the orthonormal properties of ψ. Let us multiply through Equation (6-26) by ψ^* and integrate over all space (in this case from 0 to L):

$$\int_0^L \psi_m^* f(x)\, dx = \int_0^L \psi_m^* \sum_n a_n \psi_n \, dx = \int_0^L \sum_n a_n \psi_m^* \psi_n \, dx$$

But the integral of a sum is the sum of the integrals, so

$$\int_0^L \sum_n a_n \psi_m^* \psi_n \, dx = \sum_n a_n \int_0^L \psi_m^* \psi_n \, dx$$

Every integral in the sum will vanish, except when $m = n$, in which case the integral will be unity. Thus, we can write

$$a_n = \int_0^L \psi_n^* f(x)\, dx = \sqrt{\frac{2}{L}} \int_0^L f(x) \sin \frac{n\pi x}{L} dx \tag{6-27}$$

For the case $f(x) = 3x$, Equation (6-27) is

$$a_n = 3\sqrt{\frac{2}{L}}\int_0^L x \sin\frac{n\pi x}{L}\,dx$$

which integrates to give

$$a_n = -3\sqrt{\frac{2}{L}}\left(\frac{L^2}{n\pi}\right)(-1)^n$$

With these coefficients, the Fourier series expansion becomes

$$f(x) = \sum_n -\frac{6L}{n\pi}(-1)^n \sin\frac{n\pi x}{L}$$

$$f(x) = \frac{6L}{\pi}\left(\sin\frac{\pi x}{L} - \frac{1}{2}\sin\frac{2\pi x}{L} + \frac{1}{3}\sin\frac{3\pi x}{L} - + \ldots\right)$$

A graph of the first five terms of this series, along with a graph of the original function (the dashed line) between $x = 0$ and $x = L$ is shown in Fig. 6-1. Since the function is plotted in fractional units of L, the graph actually is from $x = 0$ to $x = 1$. Note that even with only five terms in the Fourier series, the series does a pretty fair job of following the actual function. Of course, more terms in the series will give a better fit. The series does struggle to fit the function as x approaches L, since the actual function is discontinuous at that point, while the series goes to zero. This behavior, however, is normal when one attempts to approximate a nonperiodic, discontinuous function with a sum of sine waves.

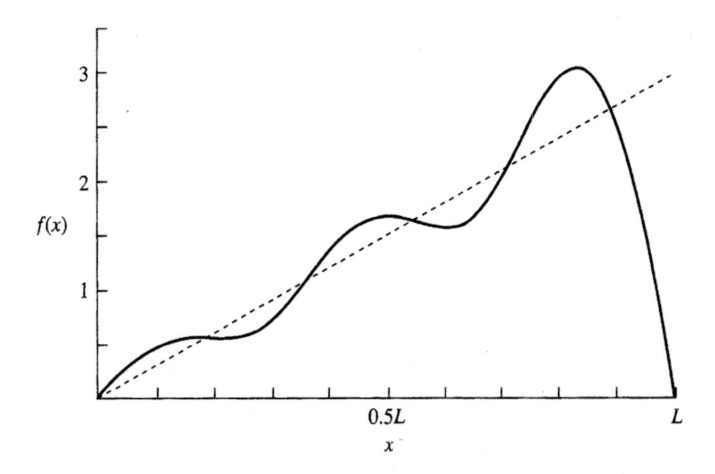

Figure 6-1 Graph of first five terms of the Fourier series expansion of the function $f(x) = 3x$, along with graph of the actual function (dashed line).

SUGGESTED READINGS

1. BRADLEY, GERALD L., and SMITH, KARL J. *Calculus*. Upper Saddle River, NJ: Prentice Hall, 1995.
2. NAGLE, R. KENT, and SAFF, EDWARD B. *Fundamentals of Differential Equations and Boundary Value Problems*, 2d ed. Boston: Addison-Wesley, 1996.
3. VARBERG, DALE, and PURCELL, EDWIN J. *Calculus*, 7th ed. Upper Saddle River, NJ: Prentice Hall, 1997.

PROBLEMS

1. For each of the following series, determine whether the series is convergent or divergent, and if it is convergent, determine the sum:

(a) $1 + 3 + 5 + 7 + 9 + \ldots$

(b) $\frac{3}{2} + \frac{3}{4} + \frac{3}{6} + \frac{3}{8} + \frac{3}{10} + \ldots$

(c) $4 + \frac{4}{3} + \frac{4}{9} + \frac{4}{27} + \frac{4}{81} + \ldots$

(d) $\frac{1}{1!} + \frac{1}{2!} + \frac{1}{3!} + \frac{1}{4!} + \frac{1}{5!} + \ldots$

(e) $\frac{1}{1} - \frac{1}{2} + \frac{1}{3} - \frac{1}{4} + \frac{1}{5} - + \ldots$

(f) $\frac{1}{4} + \frac{2}{7} + \frac{3}{10} + \frac{4}{13} + \frac{5}{16} + \ldots$

(g) $1 + \frac{1}{4} + \frac{1}{9} + \frac{1}{16} + \frac{1}{25} + \ldots$

(h) $1 + \frac{1}{\sqrt{2}} + \frac{1}{\sqrt{3}} + \frac{1}{\sqrt{4}} + \frac{1}{\sqrt{5}} + \ldots$

(i) $\frac{1}{2} + \frac{2!}{2^2} + \frac{3!}{2^3} + \frac{4!}{2^4} + \frac{5!}{2^5} + \ldots$

(j) $1 - \frac{1}{2} + \frac{1}{4} - \frac{1}{8} + \frac{1}{16} - + \ldots$

2. Use the ratio test to determine whether the following series are convergent or divergent:

(a) $\frac{1}{2} + \frac{1}{2^2} + \frac{1}{2^3} + \frac{1}{2^4} + \frac{1}{2^5} + \ldots$

(b) $3 + \frac{3^2}{2} + \frac{3^3}{3} + \frac{3^4}{4} + \frac{3^5}{5} + \ldots$

(c) $\frac{1}{2} + \frac{2}{3} + \frac{3}{4} + \frac{4}{5} + \frac{5}{6} + \ldots$

(d) $2 + \frac{2^2}{2^2} + \frac{2^3}{3^2} + \frac{2^4}{4^2} + \frac{2^5}{5^2} + \ldots$

(e) $\frac{1}{2} + \frac{2^2}{2^2} + \frac{3^2}{2^3} + \frac{4^2}{2^4} + \frac{5^2}{2^5} + \ldots$

(f) $3 + \frac{3^2}{2!} + \frac{3^3}{3!} + \frac{3^4}{4!} + \frac{3^5}{5!} + \ldots$

(g) $1 + \frac{1!}{2} + \frac{2!}{3} + \frac{3!}{4} + \frac{4!}{5} + \ldots$

(h) $\frac{1}{3!} + \frac{1}{6!} + \frac{1}{9!} + \frac{1}{12!} + \frac{1}{15!} + \ldots$

(i) $1 + \frac{1}{2} + \frac{2!}{2^2} + \frac{3!}{2^3} + \frac{4!}{2^4} + \ldots$

(j) $\frac{1}{3} + \frac{2^2}{4} + \frac{3^2}{5} + \frac{4^2}{6} + \frac{5^2}{7} + \ldots$

3. Determine the interval of convergence for the following power series:

(a) $1 + x + x^2 + x^3 + \ldots$

(b) $1 - 2x + 3x^2 - 4x^3 + - \ldots$

(c) $1 + x + \frac{x^2}{2!} + \frac{x^3}{3!} + \ldots$

(d) $x - \frac{x^2}{2} + \frac{x^3}{3} - \frac{x^4}{4} + - \ldots$

(e) $1 - x^2 + \frac{x^4}{2!} - \frac{x^6}{3!} + - \ldots$

(f) $x - \frac{1}{3}x^3 + \frac{1}{5}x^5 - \frac{1}{7}x^7 + - \ldots$

(g) $x - \frac{x^3}{3!} + \frac{x^5}{5!} - \frac{x^7}{7!} + - \ldots$

(h) $(x-1) - \frac{1}{2}(x-1)^2 + \frac{1}{3}(x-1)^3 - \frac{1}{4}(x-1)^4 + - \ldots$

(i) $1 + \frac{x}{2} + \frac{x^2}{4} + \frac{x^3}{8} + \frac{x^4}{16} + \ldots$

(j) $1 + (x+2) + (x+2)^2 + (x+2)^3 + \ldots$

4. Expand the following functions in a Maclaurin series:

(a) $\frac{1}{1+x}$

(b) $\frac{1}{(1+x)^2}$

(c) $(1+x)^{1/2}$

(d) $\ln(1-x)$

(e) e^{-x^2}

(f) a^x

(g) $\cos x$

(h) $(1 + x)^3$

5. Show that, for small values of X_B, $\ln(1 - X_B) \cong -X_B$.
6. Show that, for small values of θ, $\sin\theta \cong \theta$.
7. Show, by expanding $\sin x$ in powers of $(x - a)$, that the series converges most rapidly as x approaches a.
8. Evaluate the integral $\int_0^4 e^{-x^2}dx$ by expanding the function in a Maclaurin series. (Use only the first eight terms.)
9. Show that the solutions $\psi_n = \sqrt{\frac{2}{L}}\sin\frac{n\pi x}{L}$ to the problem of the particle in the one-dimensional box are orthogonal and normalized.
10. Find the Fourier transform of the function

$$f(x) = \begin{cases} 0; & x < -\pi \\ x; & -\pi < x < \pi \\ 0; & x > \pi \end{cases}$$

11. Find the Fourier transform of the step function

$$f(x) = \begin{cases} 0; & x < -L \\ \frac{\sqrt{2\pi}}{2L}; & -L < x < +L \\ 0; & x > L \end{cases}$$

12. Of the Fourier transforms that occur throughout mathematics, chemistry, and physics, perhaps the most striking are those which occur in the area of diffraction. For example, we can use a Fourier transform to reorganize the information found in an X-ray diffraction pattern (in effect, we are doing the job of a lens) and transform the information back into an "image" of a crystal. Consider the following one-dimensional crystal structure problem: Let

$$F_0 = +52.0 \qquad F_3 = +25.8$$

$$F_1 = -20.0 \qquad F_4 = -8.9$$

$$F_2 = -14.5 \qquad F_5 = -7.2$$

For a centrosymmetric wave (i.e., a wave that is symmetric about the region of space in which it exists), the Fourier series is

$$f(x) = F_0 + \sum_n F_n \cos 2\pi nx \qquad (n = 1 \text{ to } \infty)$$

where F_j represents the preceding Fourier coefficients (called *structure factors*). Plot $f(x)$ from $x = 0$ to $x = 1$ in steps of 0.05, and show that the first six terms approximate a one-dimensional unit cell containing two atoms, one at $x = \frac{1}{3}$ and the other at $x = \frac{2}{3}$. This problem is best solved with a spreadsheet, such as Excel. (See Chapter 11.)

13. The step function

$$f(x) = \begin{cases} -1; & -\pi \leq x < 0 \\ +1; & 0 < x \leq +\pi \end{cases}$$

can be described by the Fourier series $f(x) = \sum_n a_n \sin nx$, where

$$a_n = \frac{2}{\pi} \int_0^\pi f(x) \sin nx \, dx$$

Plot the actual function from $-\pi$ to $+\pi$, and compare the result with a plot of the Fourier series containing the first five nonzero terms. This problem is more easily solved with a spreadsheet, such as Excel. (See Chapter 11.)

14. Find the Fourier transform of the function

$$f(t) = \begin{cases} 0; & t < -\tau/2 \\ \sqrt{\frac{\pi}{2}} \sin \omega_0 t; & -\tau/2 < t < +\tau/2 \\ 0; & t > +\tau/2 \end{cases}$$

where the sine wave exists only in the period τ.

CHAPTER 7

Differential Equations

7-1 INTRODUCTION

The equation

$$f\left(x, y, \frac{dy}{dx}, \frac{d^2y}{dx^2}, \frac{d^3y}{dx^3}, \ldots, \frac{d^ny}{dx^n}\right) = 0 \qquad (7\text{-}1)$$

where $y = f(x)$, is known as a *differential equation*. The *order* of the differential equation is the order of the highest derivative that appears in it. Hence,

$$\frac{d^3y}{dx^3} + y^6 = 0$$

is an example of a third-order differential equation.

A linear differential equation is a differential equation having the form

$$A_0(x)\frac{d^ny}{dx^n} + A_1(x)\frac{d^{n-1}y}{dx^{n-1}} + \ldots + A_{n-1}(x)\frac{dy}{dx} + A_n(x)y + B(x) = 0 \qquad (7\text{-}2)$$

where $A_0(x) \neq 0$. A third-order linear differential equation, therefore, has the form

$$A_0(x)\frac{d^3y}{dx^3} + A_1(x)\frac{d^2y}{dx^2} + A_2(x)\frac{dy}{dx} + A_3(x)y + B(x) = 0$$

It is customary to divide through the equation by $A_0(x)$, giving

$$\frac{A_0(x)}{A_0(x)}\frac{d^3y}{dx^3} + \frac{A_1(x)}{A_0(x)}\frac{d^2y}{dx^2} + \frac{A_2(x)}{A_0(x)}\frac{dy}{dx} + \frac{A_3(x)}{A_0(x)}y + \frac{B(x)}{A_0(x)} = 0 \qquad (7\text{-}3)$$

or

$$\frac{d^3y}{dx^3} + N(x)\frac{d^2y}{dx^2} + P(x)\frac{dy}{dx} + Q(x)y + R(x) = 0 \qquad (7\text{-}4)$$

When the variable $R(x)$ equals zero, Equation (7-4) is known as a *reduced equation*:

$$\frac{d^3y}{dx^3} + N(x)\frac{d^2y}{dx^2} + P(x)\frac{dy}{dx} + Q(x)y = 0 \tag{7-5}$$

If a function and its derivatives are substituted into Equation (7-1) and satisfy the equation, then the function is said to be a *solution* of the differential equation.

7-2 LINEAR COMBINATIONS

Suppose that $u(x)$ and $v(x)$ are two solutions of a linear differential equation, such as Equation (7-5). We easily can show, then, that a *linear combination* of the two solutions, namely,

$$\phi = c_1u(x) + c_2v(x)$$

where c_1 and c_2 are arbitrary constants, also is a solution of Equation (7-5). If we substitute ϕ and its derivatives into Equation (7-5), we obtain the equation

$$c_1\frac{d^3u}{dx^3} + c_2\frac{d^3v}{dx^2} + c_1N(x)\frac{d^2u}{dx^2} + c_2N(x)\frac{d^2v}{dx^2}$$
$$+ c_1P(x)\frac{du}{dx} + c_2P(x)\frac{dv}{dx} + c_1Q(x)u + c_2Q(x)v = 0$$

Collecting terms gives

$$\begin{aligned} &c_1\left(\frac{d^3u}{dx^3} + N(x)\frac{d^2u}{dx^2} + P(x)\frac{du}{dx} + Q(x)u\right) \\ &+c_2\left(\frac{d^3v}{dx^3} + N(x)\frac{d^2v}{dx^2} + P(x)\frac{dv}{dx} + Q(x)v\right) = 0 \end{aligned} \tag{7-6}$$

However, since both $u(x)$ and $v(x)$ are solutions of Equation (7-5), each term in the parentheses in Equation (7-6) is identically equal to zero and the equation is identically satisfied. Thus, ϕ also is a solution of the equation.

In general, then, if $u_1(x), u_2(x), u_3(x), \ldots$ are solutions of a linear differential equation then a linear combination of these solutions, $\phi = c_1u_1 + c_2u_2 + c_3u_3 + \ldots$, also is a solution of the equation.

7-3 FIRST-ORDER DIFFERENTIAL EQUATIONS

A *first-order linear differential equation* is a differential equation having the general form

$$\frac{dy}{dx} + Q(x)y + R(x) = 0 \tag{7-7}$$

The reduced form of this equation can be solved by simple integration, using a technique known as *separation of variables*. To separate variables, we put all the y variables on one side of the equation and all the x variables on the other side:

$$\frac{dy}{dx} + Q(x)y = 0$$

$$\frac{dy}{y} = -Q(x)\,dx$$

$$\int \frac{dy}{y} = -\int Q(x)\,dx$$

$$\ln y = -\int Q(x)\,dx + C$$

Taking the antilogarithm of the last equation gives

$$y = Ae^{-\int Q(x)\,dx} \tag{7-8}$$

where $A = e^C$ is a constant.

Examples

1. The rate of a certain chemical reaction is found to be proportional to the concentration of the reactant at any time t. Find the integrated rate equation describing such a process.

   Solution. The rate of a reaction can be described by the derivative $-d(A)/dt$, which gives the rate of decrease of the concentration of the reactant A. Therefore,

   $$-\frac{d(A)}{dt} = k(A)$$

   where k is a constant of proportionality. Separating the variables gives

   $$\int \frac{d(A)}{(A)} = -\int k dt = -k \int dt$$

   $$\ln(A) = -kt + C$$

We can evaluate the constant of integration by assuming that at $t = 0$, $(A) = (A)_0$, some initial concentration of A. This results in $C = \ln(A)_0$. Therefore,

$$\ln(A) = -kt + \ln(A)_0$$

or

$$(A) = (A)_0 e^{-kt}$$

which describes an exponential decay.

2. The slope of a graph of pressure versus temperature describing any phase change of a pure substance is given by the Clapeyron equation

$$\frac{dP}{dT} = \frac{\Delta H}{T(V_f - V_i)}$$

where ΔH is the enthalpy change attending the phase change and where V_i and V_f are the molar volumes of the initial and final phases, respectively. Find the integrated form of this equation for the vaporization of a liquid, assuming the vapor phase to be a perfect gas.

Solution. For a liquid–vapor phase change, the Clapeyron equation becomes

$$\frac{dP}{dT} = \frac{\Delta H_{\text{vap}}}{T(V_g - V_l)}$$

where P is the vapor pressure of the liquid at any temperature T. We assume that V_g is much larger than V_l, which allows us to drop V_l from the equation, and since the vapor is assumed to be a perfect gas, it follows that $V_g = RT/P$. Substituting this expression for V_g back into the equation yields

$$\frac{dP}{dT} = \frac{P\Delta H_{\text{vap}}}{RT^2}$$

Separating variables and assuming that ΔH_{vap} is constant, we have

$$\int \frac{dP}{P} = \frac{\Delta H_{\text{vap}}}{R} \int \frac{dT}{T^2}$$

which integrates to give the Clausius–Clapeyron equation

$$\ln P = -\frac{\Delta H_{\text{vap}}}{RT} + C$$

Let us now consider the nonreduced equation

$$\frac{dy}{dx} + Q(x)y = f(x) \tag{7-9}$$

Equation (7-8) can help us solve Equation (7-9). We observe that

$$\frac{d}{dx}[e^{\int Q(x)\,dx}y] = e^{\int Q(x)\,dx}\frac{dy}{dx} + Q(x)ye^{\int Q(x)\,dx} \tag{7-10}$$

Therefore, if we multiply Equation (7-9) through by $e^{\int Q(x)\,dx}$, we have

$$e^{\int Q(x)\,dx}\frac{dy}{dx} + e^{\int Q(x)\,dx}Q(x)y = e^{\int Q(x)\,dx}f(x) \tag{7-11}$$

which can now be integrated. The integral of the left side of Equation (7-11) is found in Equation (7-10). Hence, we have

$$e^{\int Q(x)\,dx}y = \int e^{\int Q(x)\,dx}f(x)\,dx + C \tag{7-12}$$

The term $e^{\int Q(x)\,dx}$ is known as an *integrating factor.*

Example

In the consecutive reaction $A \xrightarrow{k_1} B \xrightarrow{k_2} C$, where k_1 and k_2 are rate constants, the concentration of B, (B), follows the rate equation

$$\frac{d(B)}{dt} = k_1(A)_0e^{-k_1t} - k_2(B)$$

where $(A)_0$ represents the initial concentration of A. Find the integrated rate equation describing the concentration of B as a function of time.

Solution. We first put the rate equation in the form given by Equation (7-9):

$$\frac{d(B)}{dt} + k_2(B) = k_1(A)_0e^{-k_1t}$$

The integrating factor in this case is $e^{\int k_2\,dt} = e^{k_2t}$. Multiplying through by this integrating factor gives

$$e^{k_2t}\frac{d(B)}{dt} + k_2(B)e^{k_2t} = k_1(A)_0e^{k_2t}e^{-k_1t}$$

The left side of this equation integrates to $e^{k_2 t}(B)$. The right side of the equation is easily integrated if we combine the exponentials. We have

$$e^{k_2 t}(B) = k_1(A)_0 \int e^{(k_2-k_1)t}\,dt$$

which gives

$$e^{k_2 t}(B) = \frac{k_1(A)_0}{(k_2 - k_1)} e^{(k_2-k_1)t} + C$$

Again, the constant of integration can be determined by assuming that $(B) = 0$ at $t = 0$. With this assumption and some algebraic manipulation, we obtain the final equation by separating the exponential on the right side of the equation into two exponentials and dividing through by $e^{k_2 t}$:

$$(B) = \frac{k_1(A)_0}{(k_2 - k_1)}(e^{-k_1 t} - e^{-k_2 t})$$

7-4 SECOND-ORDER DIFFERENTIAL EQUATIONS WITH CONSTANT COEFFICIENTS

One type of linear differential equation that is extremely important to the study of physical chemistry, and indeed to the physical sciences as a whole, is the *second-order linear differential equation with constant coefficients*, having the general form

$$\frac{d^2y}{dx^2} + A\frac{dy}{dx} + By = R(x) \qquad (7\text{-}13)$$

where A and B are constants. Consider, first, the reduced equation, with $R(x) = 0$:

$$\frac{d^2y}{dx^2} + A\frac{dy}{dx} + By = 0 \qquad (7\text{-}14)$$

A trivial solution of this equation—and, in fact, of all reduced equations, regardless of their order—is $y = 0$. Let us guess that a nontrivial solution is[1]

$$y = e^{mx} \neq 0 \qquad (7\text{-}15)$$

[1]One acceptable way to solve a differential equation is to guess a solution (called an *Ansatz* in German) and to substitute it and its derivatives into the differential equation to see whether it satisfies the equation.

where m is a constant. Taking the first and second derivatives of this equation, we have

$$\frac{dy}{dx} = me^{mx} \quad \text{and} \quad \frac{d^2y}{dx^2} = m^2e^{mx}$$

Substituting these back into Equation (7-14) gives

$$m^2 e^{mx} + Ame^{mx} + Be^{mx} = 0$$

Since $e^{mx} \neq 0$, we can divide through by e^{mx}, giving

$$m^2 + Am + B = 0 \tag{7-16}$$

Equation (7-16) is called the *auxiliary equation.* Hence, $y = e^{mx}$ is a solution of Equation (7-14), provided that there is an m such that

$$m = \frac{-A \pm \sqrt{A^2 - 4B}}{2} \tag{7-17}$$

Depending on the sign or magnitude (or both) of the constants A and B, the roots of Equation (7-17) will be real, imaginary, or complex. Let us consider the solution of the differential equation in each of these cases.

Real Roots Consider, first, the case in which the roots of the auxiliary equation are real—say, $m = \pm a$. Thus,

$$y = e^{ax} \quad \text{and} \quad y = e^{-ax}$$

are each solutions of Equation (7-14), called *particular solutions.* A general solution of Equation (7-14) when the roots of the auxiliary equation are real is found by taking a linear combination of the two particular solutions—that is,

$$y = c_1e^{ax} + c_2e^{-ax} \tag{7-18}$$

where c_1 and c_2 are arbitrary constants. We see, then, that when the roots of the auxiliary equation are real, the solution of the differential equation is real and is the combination of an exponential increase plus an exponential decay.

Imaginary Roots Consider, next, the case in which the roots of Equation (7-17) are purely imaginary, say, $m = \pm ib$, where $i = \sqrt{-1}$. (See Chapter 1.) Thus,

$$y = e^{ibx} \quad \text{and} \quad y = e^{-ibx}$$

are two particular solutions of the equation. Again, a general solution is found by taking a linear combination of the particular solutions:

$$y = c_1 e^{ibx} + c_2 e^{-ibx} \tag{7-19}$$

Recall from Chapter 1 that the exponentials $e^{i\theta}$ and $e^{-i\theta}$ can be related to sine and cosine functions through Euler's relations. Without proof, then, another form of the general solution of Equation (7-14), this time when the roots of the auxiliary equation are imaginary, is

$$y = c_1' \sin bx + c_2' \cos bx \tag{7-20}$$

Since sine and cosine functions describe waves, it is not surprising that solutions of wave equations have the form given by Equations (7-19) and (7-20).

Complex Roots Consider, now, the case in which the roots are complex, say, $m = a \pm ib$. In this case, we have the two particular solutions

$$y_1 = e^{(a+ib)x} \quad \text{and} \quad y_2 = e^{(a-ib)x}$$

A general solution is, therefore,

$$\begin{aligned} y &= c_1 e^{(a+ib)x} + c_2 e^{(a-ib)x} \\ &= c_1 e^{ax} e^{ibx} + c_2 e^{ax} e^{-ibx} \\ &= e^{ax}(c_1 e^{ibx} + c_2 e^{-ibx}) \end{aligned}$$

Double Roots There is a possibility that the two roots of the auxiliary equation may be real and equal. In this case, the general solution is (again, without proof)

$$y = c_1 e^{ax} + c_2 x e^{ax} \tag{7-21}$$

Examples

1. Solve the equation $\dfrac{d^2y}{dx^2} - 4\dfrac{dy}{dx} + 4y = 0.$

   To find the auxiliary equation, substitute $y = e^{mx}$ and its derivatives into the given equation. This results in

$$\begin{aligned} m^2 - 4m + 4 &= 0 \\ (m - 2)^2 &= 0 \end{aligned}$$

which has the real, double roots $m = 2, 2$. The general solution of the equation is then

$$y = c_1 e^{2x} + c_2 x e^{2x}$$

2. Solve $\dfrac{d^2y}{dx^2} + 5y = 0$.

Substituting $y = e^{mx}$ and its derivatives into this equation gives

$$m^2 + 5 = 0$$

which has the imaginary roots $m = \pm i\sqrt{5}$. The general solution of this equation is, accordingly,

$$y = c_1 e^{+i\sqrt{5}x} + c_2 e^{-i\sqrt{5}x}$$

or, in sine–cosine form,

$$y = c_1' \sin \sqrt{5}x + c_2' \cos \sqrt{5}x$$

3. Solve the Schrödinger equation

$$\frac{d^2\psi}{dx^2} + \frac{8\pi^2 mE}{h^2}\psi = 0$$

where m, E, and h are constants.

Let $\psi = e^{kx}$. Substituting this equation and its derivatives into the Schrödinger equation, we have

$$k^2 + \frac{8\pi^2 mE}{h^2} = 0$$

which has the imaginary roots $k = +i\sqrt{\dfrac{8\pi^2 mE}{h^2}}$ and $k = -i\sqrt{\dfrac{8\pi^2 mE}{h^2}}$. Choosing the sine–cosine form of the solution, we obtain

$$\psi = A \sin \sqrt{\frac{8\pi^2 mE}{h^2}}x + B \cos \sqrt{\frac{8\pi^2 mE}{h^2}}x$$

where A and B are constants.

7-5 GENERAL SERIES METHOD OF SOLUTION

In certain cases, differential equations cannot be solved by the simple method outlined in Section 7-4. For that reason, consider another important method for solving differential equations, called the *series method* of solution. Generally, this method is used for equations that are not reduced or for equations whose coefficients are not constant. To see how the series method works, let us introduce it on a very simple reduced equation.

Consider the equation

$$\frac{d^2y}{dx^2} + \beta^2 y = 0 \tag{7-22}$$

where β is a constant. We easily can show, using the method outlined in Section 7-4, that two particular solutions of Equation (7-22) are $\sin \beta x$ and $\cos \beta x$. Let us assume, however, that the solution of Equation (7-22) is a power series of the form (see Chapter 6)

$$y = a_\kappa x^\kappa + a_{\kappa+1} x^{(\kappa+1)} + a_{\kappa+2} x^{(\kappa+2)} + \dots = \sum_{n=0}^{\infty} a_{\kappa+n} x^{(\kappa+n)} \tag{7-23}$$

where κ represents the lowest power that x can have in the summation. Taking the first and second derivatives of this equation gives

$$\frac{dy}{dx} = \sum_{n=1}^{\infty} (\kappa + n) a_{\kappa+n} x^{(\kappa+n-1)}$$

and

$$\frac{d^2y}{dx^2} = \sum_{n=2}^{\infty} (\kappa + n - 1)(\kappa + n) a_{\kappa+n} x^{(\kappa+n-2)}$$

Substituting the second derivative and y back into Equation (7-22), we have

$$\sum_{n=2}^{\infty} (\kappa + n - 1)(\kappa + n) a_{\kappa+n} x^{(\kappa+n-2)} + \beta^2 \sum_{n=0}^{\infty} a_{\kappa+n} x^{(\kappa+n)} = 0 \tag{7-24}$$

Equation (7-24) must hold for every value of x, and that will be true only if every coefficient of the power of x is identically equal to zero. Since n cannot be negative, the lowest power of x in the first summation in Equation (7-24) is $x^{(\kappa-2)}$, where $n = 0$. Accordingly, substituting this into the first summation gives

$$(\kappa - 1)\kappa a_\kappa = 0 \tag{7-25}$$

Equation (7-25) is called an *indicial equation.* Since $a_k \neq 0$, $(\kappa - 1)\kappa = 0$, which gives

$$\kappa = 0, 1$$

We can combine the summations in Equation (7-24) if we replace n in the first summation with $n + 2$, giving[2]

$$\sum_{n=0}^{\infty}(\kappa + n + 1)(\kappa + n + 2)a_{\kappa+n+2}x^{(\kappa+n)} + \beta^2\sum_{n=0}^{\infty}a_{\kappa+n}x^{(\kappa+n)} = 0$$

$$\sum_{n=0}^{\infty}\{(\kappa + n + 1)(\kappa + n + 2)a_{\kappa+n+2} + \beta^2 a_{\kappa+n}\}x^{(\kappa+n)} = 0$$

Since $x^{\kappa+n} \neq 0$, the term in braces must equal zero, giving

$$a_{\kappa+n+2} = \frac{-\beta^2 a_{\kappa+n}}{(\kappa + n + 1)(\kappa + n + 2)} \tag{7-26}$$

Equation (7-26) is called a *recursion equation* or *recursion formula*; it connects the coefficients of the series. We see in this case that we have two series expansions: one for which $\kappa = 0$ and one for which $\kappa = 1$. Let us look at the $\kappa = 0$ expansion. Under these circumstances,

$$a_{n+2} = \frac{-\beta^2 a_n}{(n + 1)(n + 2)}$$

$$a_2 = \frac{-\beta^2 a_0}{2}, a_4 = \frac{\beta^4 a_2}{24}, \ldots$$

With $a_0 = 1$, we have

$$y = 1 - \frac{\beta^2}{2}x^2 + \frac{\beta^4}{24}x^4 - + \ldots = 1 - \frac{(\beta x)^2}{2} + \frac{(\beta x)^4}{24} - + \ldots$$

which is just the series expansion for cos βx, a particular solution of Equation (7-22). Substituting $\kappa = 1$ into Equation (7-26) and letting $a_1 = \beta$, we find it easy to show that the series which results is the series expansion for sin βx, another particular solution of Equation (7-22).

[2]Adding two more terms to an infinite number of terms is the same as adding zero to those terms.

7-6 SPECIAL POLYNOMIAL SOLUTIONS OF DIFFERENTIAL EQUATIONS

Hermite's Equation Consider the differential equation

$$\frac{d^2y}{dx^2} - 2x\frac{dy}{dx} + 2\alpha y = 0 \tag{7-27}$$

Equation (7-27) is known as *Hermite's equation.* The roots of the indicial equation in this case are $\kappa = 0, 1$ (see Section 7-5), which give rise to the recursion formula

$$a_{n+2} = \frac{2(\kappa + n) - 2\alpha}{(\kappa + n + 2)(k + n + 1)}a_n$$

For $\kappa = 0$, we have

$$y = a_0\left(1 - \frac{2\alpha}{2!}x^2 + \frac{2^2\alpha(\alpha - 2)}{4!}x^4 - \frac{2^3\alpha(\alpha - 2)(\alpha - 4)}{6!}x^6 + - \ldots\right)$$

For $\kappa = 1$, we have

$$y = a_0x\left(1 - \frac{2(\alpha - 1)}{3!}x^2 + \frac{2^2(\alpha - 1)(\alpha - 3)}{5!}x^4 - + \ldots\right)$$

The general solution of Equation (7-27) is the superposition of these two particular solutions. The appropriate choice of

$$a_0 = (-1)^{n/2}\frac{n!}{(\frac{n}{2})!}$$

leads to a set of solutions known as the *Hermite polynomials* of degree n:

$$\begin{aligned} H_n(x) = (2x)^n &- \frac{n(n-1)}{1!}(2x)^{n-2} \\ &+ \frac{n(n-1)(n-2)(n-3)}{2!}(2x)^{n-4} - \ldots \end{aligned} \tag{7-28}$$

Example

The Schrödinger equation describing a simple one-dimensional harmonic oscillator is

$$\frac{d^2\psi}{dx^2} + \frac{8\pi^2 m}{h^2}\left(E - \frac{1}{2}kx^2\right)\psi = 0 \tag{7-29}$$

where π, m, h, E, and k are constants. Show that this differential equation can be solved by using Hermite polynomials.

Solution. Let

$$\varepsilon = \frac{8\pi^2 mE}{h^2} \quad \text{and} \quad \beta^2 = \frac{4\pi^2 mk}{h^2}$$

Substituting these equations back into Equation (7-29) gives

$$\frac{d^2\psi}{dx^2} + (\varepsilon - \beta^2 x^2)\psi = 0$$

Next, we make a change of variables by letting $\xi = \sqrt{\beta}x$:

$$\frac{d^2\psi}{dx^2} = \left(\frac{\partial^2\psi}{\partial\xi^2}\right)\left(\frac{\partial^2\xi}{\partial x^2}\right) = \sqrt{\beta}\left(\frac{\partial^2\psi}{\partial\xi^2}\right)$$

This gives

$$\frac{d^2\psi}{d\xi^2} + \left[1 - \xi^2 + \left(\frac{\varepsilon}{\beta} - 1\right)\right]\psi = 0 \tag{7-30}$$

We now let $\psi(\xi) = e^{-\xi^2/2}y(\xi)$. Taking the second derivative of this function [see Chapter 4, Section 4-2, Item 11, Example (d)] and substituting it and ψ into Equation (7-30) gives

$$\frac{d^2y}{d\xi^2} - 2\xi\frac{dy}{d\xi} + \left(\frac{\varepsilon}{\beta} - 1\right)y = 0$$

which is just Equation (7-27), Hermite's equation, the solutions of which are the Hermite polynomials.

Laguerre's Equation Consider the differential equation

$$x\frac{d^2y}{dx^2} + (1 - x)\frac{dy}{dx} + \alpha y = 0 \tag{7-31}$$

where α is a constant. This equation is known as *Laguerre's equation.* The indicial equation for Laguerre's equation has a single root $\kappa = 0$, which gives rise to the recursion formula

$$a_{n+1} = \frac{n - \alpha}{(n + 1)^2}a_n \tag{7-32}$$

The series solution is therefore

$$y = a_0\left(1 - \alpha x + \frac{\alpha(\alpha - 1)}{(2!)^2}x^2 - \dots\right) \tag{7-33}$$

Again, the choice of

$$a_0 = (-1)^n n!$$

leads to the *Laguerre polynomials* of degree n:

$$L_n(x) = (-1)^n\left(x^n - \frac{n^2}{1!}x^{n-1} + \frac{n^2(n - 1)^2}{2!}x^{n-2} + \dots + (-1)^n n!\right) \tag{7-34}$$

A differential equation closely related to Laguerre's equation is the equation

$$x\frac{d^2y}{dx^2} + (k + 1 - x)\frac{dy}{dx} + (\alpha - k)y = 0 \tag{7-35}$$

where k is an integer greater than or equal to zero. This equation is produced when Equation (7-31) is differentiated k times and y is replaced by the kth derivative. Solutions of the equation are usually represented as

$$y = \frac{d^k}{dx^k}L_n(x) \equiv L_n^k(x)$$

and are called the *associated Laguerre polynomials* of degree $(n - k)$.

A third form of the Laguerre equation, important in the wave-mechanical solution of the radial part of the hydrogen atom, is the equation

$$x\frac{d^2y}{dx^2} + 2\frac{dy}{dx} + \left[n - \frac{k-1}{2} - \frac{x}{4} - \frac{k^2-1}{4x}\right]y = 0 \tag{7-36}$$

This equation can be transformed into one having the same form as Equation (7-35) by letting $y = e^{-x/2}x^{(k-1)/2}v$. Substituting y and its derivatives into Equation (7-36) yields the equation

$$x\frac{d^2v}{dx^2} + (k + 1 - x)\frac{dv}{dx} + (n - k)v = 0 \tag{7-37}$$

which we see has exactly the same form as Equation (7-35). Thus, $v = L_n^k$, and a particular solution of Equation (7-36) is

$$y = e^{-x/2}x^{(k-1)/2}L_n^k \tag{7-38}$$

This function is called the *associated Laguerre function.*

Example

The radial part of the Schrödinger equation for the hydrogen atom is

$$\frac{1}{r^2}\frac{d}{dr}\left(r^2\frac{dR}{dr}\right) + \left[\frac{2\mu}{\hbar^2}\left(E + \frac{e^2}{4\pi\varepsilon_0 r}\right) - \frac{\ell(\ell+1)}{r^2}\right]R(r) = 0$$

where π, μ, $\hbar$, ε_0, and ℓ are constants. Show that the solutions of this equation are the associated Laguerre polynomials.

Solution. To transform the radial equation into a form that resembles Laguerre's equation, let us first expand the equation:

$$\frac{1}{r^2}\frac{d}{dr}\left(r^2\frac{dR}{dr}\right) + \left[\frac{2\mu E}{\hbar^2} + \frac{2\mu e^2}{4\pi\varepsilon_0 r\hbar^2} - \frac{\ell(\ell+1)}{r^2}\right]R(r) = 0$$

Next, we define two new constants

$$\alpha^2 = -\frac{2\mu E}{\hbar^2} \quad \text{and} \quad \beta = \frac{\mu e^2}{4\pi\varepsilon_0\alpha\hbar^2}$$

Substituting these into the equation, we have

$$\frac{1}{r^2}\frac{d}{dr}\left(r^2\frac{dR}{dr}\right) + \left[-\alpha^2 + \frac{2\beta\alpha}{r} - \frac{\ell(\ell+1)}{r^2}\right]R(r) = 0 \qquad (7\text{-}39)$$

We next make a transformation of variables, $\rho = 2\alpha r$, yielding

$$\frac{dR}{dr} = \frac{d\rho}{dr}\cdot\frac{dR}{d\rho} = 2\alpha\frac{dR}{d\rho} \quad \text{and} \quad \frac{d}{dr} = 2\alpha\frac{d}{d\rho}$$

Substituting these back into Equation (7-39) gives

$$\left(\frac{4\alpha^2}{\rho^2}\right)\frac{d}{d\rho}\left(\rho^2\frac{dR}{d\rho}\right) + \left(-\alpha^2 + \frac{4\beta\alpha^2}{\rho} - (4\alpha^2)\frac{\ell(\ell+1)}{\rho^2}\right)R(\rho) = 0$$

Dividing by $4\alpha^2$ results in

$$\frac{1}{\rho^2}\frac{d}{d\rho}\left(\rho^2\frac{dR}{d\rho}\right) + \left[-\frac{1}{4} + \frac{\beta}{\rho} - \frac{\ell(\ell+1)}{\rho^2}\right]R(\rho) = 0$$

Expanding the derivative $d/d\rho$ and remembering that it must be differentiated as a product yields

$$\frac{1}{\rho^2}\left(\rho^2\frac{d^2R}{d\rho^2} + 2\rho\frac{dR}{d\rho}\right) + \left[-\frac{1}{4} + \frac{\beta}{\rho} - \frac{\ell(\ell+1)}{\rho^2}\right]R(\rho) = 0$$

Finally, multiplying through by ρ, we obtain

$$\rho\frac{d^2R}{d\rho^2} + 2\frac{dR}{d\rho} + \left[-\frac{\rho}{4} + \beta - \frac{\ell(\ell+1)}{\rho}\right]R(\rho) = 0 \qquad (7\text{-}40)$$

which we see has exactly the same form as Equation (7-36). The solutions of Equation (7-40), then, are

$$R(\rho) = \rho^\ell e^{-\rho/2}L^k(\rho),\, k = 2\ell + 1$$

where $L^k(\rho)$ are the associated Laguerre polynomials.

Legendre's Equation Now consider an equation having the general form

$$(1 - x^2)\frac{d^2y}{dx^2} - 2x\frac{dy}{dx} + \ell(\ell + 1)y = 0 \tag{7-41}$$

where ℓ is a constant. This equation is known as *Legendre's equation.* Series solutions of Legendre's equation leads to the indicial equation having roots $\kappa = 0, 1$, and the recursion formula

$$a_{n+2} = \frac{(\kappa + n)(\kappa + n + 1) - \ell(\ell + 1)}{(\kappa + n + 1)(\kappa + n + 2)} a_n \tag{7-42}$$

Like Hermite's equation, we have a choice of even or odd solutions. When $\kappa = 0$, the significant series solution of Legendre's equation is

$$y = \left[1 - \frac{\ell(\ell + 1)}{2!}x^2 + \frac{\ell(\ell - 1)(\ell + 1)(\ell + 3)}{4!}x^4 + \ldots\right]a_0 \tag{7-43}$$

When $\kappa = 1$, the significant series solution is

$$y = \left[x - \frac{(\ell - 1)(\ell + 2)}{3!}x^3 + \frac{(\ell - 1)(\ell - 3)(\ell + 2)(\ell + 4)}{5!}x^5 + \ldots\right]a_1 \tag{7-44}$$

As in the previous special equations, series solutions of differential equations are of particular interest when the series converges to a polynomial. In the case of Legendre's equation, the "nuts and bolts" of finding the conditions for convergence are intricate and beyond the scope of this text. Students interested in the procedure are referred to the readings listed at the end of the chapter. Suffice it to say that the general series solution, which is a linear combination of Equations (7-43) and (7-44), reduces to a polynomial when ℓ is an even or odd, positive or negative integer, including zero. Under these conditions, the resulting polynomials P_ℓ, called the *Legendre polynomials*, have the form

$$P_\ell(x) = \frac{1\cdot3\cdot5\ldots(2\ell + 1)}{\ell!}\left\{x^\ell - \frac{\ell(\ell - 1)}{2(2\ell - 1)}x^{\ell-2} + \frac{\ell(\ell - 1)(\ell - 2)(\ell - 3)}{2\cdot4(2\ell - 1)(2\ell - 3)}x^{\ell-4} - \ldots\right\} \tag{7-45}$$

An equation closely related to Legendre's equation, and important in the solution of problems involving rotational motion, is

$$(1 - x^2)\frac{d^2y}{dx^2} - 2x\frac{dy}{dx} + \left(\ell(\ell + 1) - \frac{m^2}{1 - x^2}\right)y = 0 \tag{7-46}$$

where ℓ and m are integers. This equation is known as the *associated Legendre's equation* and has the particular solution

$$y = (1 - x^2)^{m/2}\frac{d^m}{dx^m}P_\ell(x) \tag{7-47}$$

This solution is known as the *associated Legendre function*, or the *associated spherical harmonics*, because it is related to the allowed standing waves on the surface of a sphere.

Example

The $\Theta(\theta)$ equation, an angular part of the Schrödinger equation describing the hydrogen atom, can be expressed as

$$\frac{1}{\sin\theta}\frac{d}{d\theta}\left(\sin\theta\frac{d\Theta}{d\theta}\right) + \left[\ell(\ell + 1) - \frac{m^2}{\sin^2\theta}\right]\Theta(\theta) = 0$$

where ℓ and m are integers. Show that solutions of this equation are the associated Legendre polynomials.

Solution. We can put the $\Theta(\theta)$ equation in the form of the associated Legendre's equation if we let $x = \cos\theta$. Therefore, $\sin^2\theta = 1 - \cos^2\theta = 1 - x^2$. Also,

$$\frac{d\Theta}{d\theta} = \frac{dx}{d\theta}\cdot\frac{d\Theta}{dx} = -\sin\theta\frac{d\Theta}{dx} \quad \text{and} \quad \frac{d}{d\theta} = -\sin\theta\frac{d}{dx}$$

Substituting these into the $\Theta(\theta)$ equation gives

$$\frac{d}{dx}\left((1 - x^2)\frac{d\Theta}{dx}\right) + \left[\ell(\ell + 1) - \frac{m^2}{1 - x^2}\right]\Theta(x) = 0$$

Taking the derivative *d/dx* of the product in the parentheses yields

$$(1 - x^2)\frac{d^2\Theta}{dx^2} - 2x\frac{d\Theta}{dx} + \left(\ell(\ell + 1) - \frac{m^2}{1 - x^2}\right)\Theta(x) = 0$$

which is just the associated Legendre's equation.

7-7 EXACT AND INEXACT DIFFERENTIALS

The expression $M(x, y)\,dx + N(x, y)\,dy$ is said to be an *exact differential* if there exists a function $F(x, y)$ for which

$$dF = \left(\frac{\partial F}{\partial x}\right)_y dx + \left(\frac{\partial F}{\partial y}\right)_x dy = M(x, y)\,dx + N(x, y)\,dy \qquad (7\text{-}48)$$

If $F(x, y)$ does not exist, $M(x, y)\,dx + N(x, y)\,dy$ is not exact and is called an *inexact differential.*

Euler's Test for Exactness If $M(x, y)\,dx + N(x, y)\,dy$ is exact, then

$$\left(\frac{\partial F}{\partial x}\right)_y = M(x, y) \quad \text{and} \quad \left(\frac{\partial F}{\partial y}\right)_x = N(x, y)$$

Taking the mixed second derivatives gives

$$\left(\frac{\partial^2 F}{\partial y\,\partial x}\right) = \left(\frac{\partial M}{\partial y}\right)_x \quad \text{and} \quad \left(\frac{\partial^2 F}{\partial x\,\partial y}\right) = \left(\frac{\partial N}{\partial x}\right)_y$$

However, since

$$\left(\frac{\partial^2 F}{\partial y\,\partial x}\right) = \left(\frac{\partial^2 F}{\partial x\,\partial y}\right)$$

it follows that

$$\left(\frac{\partial M}{\partial y}\right)_x = \left(\frac{\partial N}{\partial x}\right)_y \qquad (7\text{-}49)$$

is a necessary condition for exactness. Equation (7-49) is known as Euler's (read "oiler's") test for exactness.

Example

Show that the expression

$$dF = 3x^2y^3\,dx + 3y^2x^3\,dy$$

is exact.

Solution. Since $M(x, y) = 3x^2y^3$ and $N(x, y) = 3y^2x^3$, we have $\partial M/\partial y = 9x^2y^2$ and $\partial N/\partial x = 9x^2y^2$. Thus, $\partial M/\partial y = \partial N/\partial x$, Euler's test is satisfied, and the differential is exact.

The equation $(\partial F/\partial x)_y = M(x, y)$ can be written in the more general form

$$F(x, y) = \int M(x, y)\, dx + K(y)$$

where $K(y)$ is independent of the variable x. Taking the partial derivative of $F(x, y)$ with respect to y gives

$$\left(\frac{\partial F}{\partial y}\right)_x = \frac{\partial}{\partial y}\int M(x, y)\, dx + \frac{\partial K}{\partial y} = N(x, y)$$

or

$$\frac{\partial K}{\partial y} = N(x, y) - \frac{\partial}{\partial y}\int M(x, y)\, dx \qquad (7\text{-}50)$$

Integrating Equation (7-50) produces

$$K(y) = \int\left[N(x, y) - \frac{\partial}{\partial y}\int M(x, y)\, dx\right] dy$$

which gives

$$F(x, y) = \int M(x, y)\, dx + \int\left[N(x, y) - \frac{\partial}{\partial y}\int M(x, y)\, dx\right] dy \qquad (7\text{-}51)$$

Example

Show that the equation

$$dF = 3x^2y^2\, dx + 2x^3y\, dy$$

is exact, and determine $F(x, y)$.

Solution

$$M(x, y) = 3x^2y^2 \quad \text{and} \quad N(x, y) = 2x^3y$$

$$\frac{\partial M}{\partial y} = 6x^2y \quad \text{and} \quad \frac{\partial N}{\partial x} = 6x^2y$$

Euler's test is satisfied and the differential dF is exact. Now,

$$\int M(x, y)\,dx = \int 3x^2y^2\,dx = x^3y^2$$

and

$$\frac{\partial}{\partial y}\int M(x, y)\,dx = 2x^3y$$

Therefore,

$$\begin{aligned} K(y) &= \int\left[N(x, y) - \frac{\partial}{\partial y}\int M(x, y)\,dx\right]dy \\ &= \int (2x^3y - 2x^3y)\,dy = C \end{aligned}$$

which gives

$$F(x, y) = x^3y^2 + C$$

In Chapter 5, we introduced the concept of the *line integral* as representing the area under a curve (or path) taking some function $f(x)$ from x_1 to x_2. Exact and inexact differentials are directly related to line integrals and occupy a significant place in physical chemistry. If du is an exact differential, then the line integral ${}_L\int_{u_1}^{u_2} du$, which represents the sum of infinitesimal elements du taking the function u from u_1 to u_2, depends only on the limits of integration. That is, if du is exact, then

$${}_L\int_{u_1}^{u_2} du = u_2 - u_1 = \Delta u$$

When u represents a physical variable describing a system, that variable is said to be a state function, since the function depends only on the initial and final states of the system and not the path taking the system from state u_1 to state u_2.

If, however, du is an inexact differential, then ${}_L\int_{u_1}^{u_2} du = u$, which is a quantity that depends on a specific path taking u_1 to u_2. The physical variable represented by u is *not* a function of state, and unless the functional relationship between the variables x and y in du is known, the integral ${}_L\int_{u_1}^{u_2} du$ cannot be evaluated. For example, we saw in Chapter 5 that the work done by the expansion of a gas against an external pressure is a function of both the external pressure on the gas (part of the surroundings) and the volume of the gas. We can write this relationship as the total differential

$$dw = M(P_{\text{ext}}, V)dP_{\text{ext}} + N(P_{\text{ext}}, V)dV = -P_{\text{ext}}dV$$

Clearly, dw is not an exact differential, since $M(P_{\text{ext}}, V) = 0$, $N(P_{\text{ext}}, V) = -P_{\text{ext}}$, $(\partial M/\partial V) = 0$, and $(\partial N/\partial P_{\text{ext}}) = -1$. Thus, Euler's test is not satisfied. Consequently, the integral $\int dw = -\int P_{\text{ext}}\, dV$ cannot be evaluated, unless the dependence of the external pressure upon the volume of the gas is known. There is, however, no necessary relationship between the external pressure and the volume of the gas; therefore, an infinite number of paths exist taking the gas from $P_1V_1T_1$ to $P_2V_2T_2$. Each path is associated with a specific amount of work.

Another important property of exact differentials is that the integral of an exact differential around a closed path must be equal to zero. That is, if we integrate an exact differential, such as the differential of energy, dE, from some initial energy state E_1 around a closed cycle ending up again at state E_1, then

$$\int_{L\,E_1}^{E_1} dE = \oint dE = E_1 - E_1 = 0 \tag{7-52}$$

Such an integral is called a *cyclic integral*. Equation (7-52) can be thought of as the consummate test for a state function.

7-8 INTEGRATING FACTORS

In certain cases, an inexact differential can be made exact by multiplying it by a nonzero function called an *integrating factor*. For example, the differential equation

$$du = 2y^4\, dx + 4xy^3\, dy \tag{7-53}$$

is not exact. (To see this, apply Euler's test for exactness.) If we multiply Equation (7-53) by x, we obtain the equation

$$dF = 2xy^4\, dx + 4x^2y^3\, dy \tag{7-54}$$

which *is* exact. (Apply Euler's test to *this* equation.) The factor x is thus an integrating factor, since it transformed the inexact differential du into the exact differential dF. Moreover, since dF is an exact differential, Equation (7-54) is said to be an exact differential equation and can be solved for $F(x, y)$ by the method outlined in the previous section.

7-9 PARTIAL DIFFERENTIAL EQUATIONS

A partial differential equation is a differential equation containing partial derivatives and, therefore, more than one independent variable. An example of a partial differential equation is

$$\frac{\partial^2 u}{\partial x^2} + \frac{\partial^2 u}{\partial y^2} + u(x, y) = 0 \tag{7-55}$$

where $u = f(x, y)$. Since u is a function of two variables x and y, the partial differential equation cannot be solved by direct integration. Before a solution can be found, the variables x and y must be separated; but that may not always be possible. One way to separate variables is to assume that the solution is a product of functions of single variables. Let us attempt this type of solution on Equation (7-55). Let

$$u(x, y) = f(x)g(y) \tag{7-56}$$

Taking the partial derivatives of Equation (7-56), we have

$$\frac{\partial u}{\partial x} = g(y)\frac{\partial f}{\partial x} \quad \text{and} \quad \frac{\partial^2 u}{\partial x^2} = g(y)\frac{\partial^2 f}{\partial x^2}$$

$$\frac{\partial u}{\partial y} = f(x)\frac{\partial g}{\partial y} \quad \text{and} \quad \frac{\partial^2 u}{\partial y^2} = f(x)\frac{\partial^2 g}{\partial y^2}$$

Substituting these into Equation (7-55) results in

$$g(y)\frac{\partial^2 f}{\partial x^2} + f(x)\frac{\partial^2 g}{\partial y^2} + f(x)g(y) = 0$$

Dividing through by $f(x)g(y)$ gives

$$\frac{1}{f(x)}\frac{\partial^2 f}{\partial x^2} + \frac{1}{g(y)}\frac{\partial^2 g}{\partial y^2} + 1 = 0 \tag{7-57}$$

or

$$\frac{1}{f(x)}\frac{\partial^2 f}{\partial x^2} + \frac{1}{g(y)}\frac{\partial^2 g}{\partial y^2} = -1 \tag{7-58}$$

Note that the first term in Equation (7-58) is only a function of the variable x and the second term in the equation is only a function of the variable y. Hence, the two variables x and y have been separated. Note further that the two terms

$$\frac{1}{f(x)}\frac{\partial^2 f}{\partial x^2} \quad \text{and} \quad \frac{1}{g(y)}\frac{\partial^2 g}{\partial y^2}$$

must equal a constant for the following reason: If we assume that $f(x)$ is the variable and differentiate partially with respect to x, then we must hold $g(y)$ and its derivatives constant:

$$\frac{1}{f(x)}\frac{\partial^2 f}{\partial x^2} = -\frac{1}{g(y)}\frac{\partial^2 g}{\partial y^2} - 1$$

But a variable cannot equal the sum of two constants, so $\frac{1}{f(x)}\frac{\partial^2 f}{\partial x^2}$ also must be constant. (Call it k_1.) The same argument can be made for $\frac{1}{g(y)}\frac{\partial^2 g}{\partial y^2}$. (Call it k_2.) That is,

$$\frac{1}{f(x)}\frac{d^2 f}{dx^2} = k_1 \quad \text{and} \quad \frac{1}{g(y)}\frac{d^2 g}{dy^2} = k_2$$

where $k_1 + k_2 = -1$. The equations

$$\frac{1}{f(x)}\frac{d^2 f}{dx^2} - k_1 = 0 \quad \text{and} \quad \frac{1}{g(y)}\frac{d^2 g}{dy^2} - k_2 = 0$$

can be solved by methods outlined in previous sections of this chapter. Once $f(x)$ and $g(y)$ are known, the solution of Equation (7-55) is $u(x, y) = f(x)g(y)$.

PROBLEM. Separate the Schrödinger equation describing the hydrogen atom into three equations: an $R(r)$ equation, a $\Theta(\theta)$ equation, and a $\Phi(\phi)$ equation. The Schrödinger equation is

$$\frac{1}{r^2}\frac{\partial}{\partial r}\left(r^2\frac{\partial\psi}{\partial r}\right) + \frac{1}{r^2 \sin\theta}\frac{\partial}{\partial\theta}\left(\sin\theta\frac{\partial\psi}{\partial\theta}\right) + \frac{1}{r^2\sin^2\theta}\frac{\partial^2\psi}{\partial\phi^2}$$
$$+ \frac{2\mu}{\hbar^2}\left(E + \frac{e^2}{4\pi\varepsilon_0 r}\right)\psi(r,\theta,\phi) = 0 \tag{7-59}$$

where π, ε_0, $\hbar$, μ, and e are constants.

Solution. Assume a product solution $\psi(r, \theta, \phi) = R(r)\Theta(\theta)\Phi(\phi)$. Evaluating the derivatives, we have

$$\frac{\partial \psi}{\partial r} = \Theta(\theta)\Phi(\phi)\frac{\partial R}{\partial r}, \quad \frac{\partial \psi}{\partial \theta} = R(r)\Phi(\phi)\frac{\partial \Theta}{\partial \theta}, \quad \frac{\partial^2 \psi}{\partial \phi^2} = R(r)\Theta(\theta)\frac{\partial^2 \Phi}{\partial \phi^2}$$

Substituting these back into Equation (7-59) gives

$$\frac{\Theta(\theta)\Phi(\phi)}{r^2}\frac{\partial}{\partial r}\left(r^2\frac{\partial R}{\partial r}\right) + \frac{R(r)\Phi(\phi)}{r^2 \sin\theta}\frac{\partial}{\partial \theta}\left(\sin\theta\frac{\partial \Theta}{\partial \theta}\right) + \frac{R(r)\Theta(\theta)}{r^2 \sin^2\theta}\frac{\partial^2 \Phi}{\partial \phi^2}$$
$$+ \frac{2\mu}{\hbar^2}\left(E + \frac{e^2}{4\pi\varepsilon_0 r}\right)R(r)\Theta(\theta)\Phi(\phi) = 0$$

Dividing through by $\psi(r, \theta, \phi) = R(r)\Theta(\theta)\Phi(\phi)$ yields

$$\frac{1}{R(r)}\frac{1}{r^2}\frac{\partial}{\partial r}\left(r^2\frac{\partial R}{\partial r}\right) + \frac{1}{\Theta(\theta)}\frac{1}{r^2 \sin\theta}\frac{\partial}{\partial \theta}\left(\sin\theta\frac{\partial \Theta}{\partial \theta}\right)$$
$$+ \frac{1}{\Phi(\phi)}\frac{1}{r^2 \sin^2\theta}\frac{\partial^2 \Phi}{\partial \phi^2} + \frac{2\mu}{\hbar^2}\left(E + \frac{e^2}{4\pi\varepsilon_0 r}\right) = 0 \qquad (7\text{-}60)$$

Next, we multiply through Equation (7-60) by $r^2 \sin^2\theta$, to isolate the ϕ dependence:

$$\frac{\sin^2\theta}{R(r)}\frac{\partial}{\partial r}\left(r^2\frac{\partial R}{\partial r}\right) + \frac{\sin\theta}{\Theta(\theta)}\frac{\partial}{\partial \theta}\left(\sin\theta\frac{\partial \Theta}{\partial \theta}\right) + \frac{1}{\Phi(\phi)}\frac{\partial^2 \Phi}{\partial \phi^2}$$
$$+ \frac{2\mu r^2 \sin^2\theta}{\hbar^2}\left(E + \frac{e^2}{4\pi\varepsilon_0 r}\right) = 0 \qquad (7\text{-}61)$$

By previous arguments, $[1/\Phi(\phi)](\partial^2\Phi/\partial\phi^2)$ must equal a constant (call it $-m^2$):

$$\frac{1}{\Phi(\phi)}\frac{d^2\Phi}{d\phi^2} = -m^2$$

(Notice that partial derivative notation is not needed, since the equation contains only one variable.) The remainder of Equation (7-61) must equal $+m^2$:

$$\frac{\sin^2\theta}{R(r)}\frac{\partial}{\partial r}\left(r^2\frac{\partial R}{\partial r}\right) + \frac{\sin\theta}{\Theta(\theta)}\frac{\partial}{\partial \theta}\left(\sin\theta\frac{\partial \Theta}{\partial \theta}\right)$$
$$+ \frac{2\mu r^2 \sin^2\theta}{\hbar^2}\left(E + \frac{e^2}{4\pi\varepsilon_0 r}\right) = +m^2$$

Now, divide through this equation by $\sin^2 \theta$ to separate the r terms and the θ terms:

$$\frac{1}{R(r)}\frac{\partial}{\partial r}\left(r^2\frac{\partial R}{\partial r}\right) + \frac{1}{\Theta(\theta)}\frac{1}{\sin\theta}\frac{\partial}{\partial\theta}\left(\sin\theta\frac{\partial\Theta}{\partial\theta}\right) + \frac{2\mu r^2}{\hbar^2}\left(E + \frac{e^2}{4\pi\varepsilon_o r}\right) - \frac{m^2}{\sin^2\theta} = 0$$

The combined r terms must equal a constant [call it $\ell(\ell + 1)$], and the combined θ terms must therefore equal $-\ell(\ell + 1)$:

$$\frac{1}{R(r)}\frac{d}{dr}\left(r^2\frac{dR}{dr}\right) + \frac{2\mu r^2}{\hbar^2}\left(E + \frac{e^2}{4\pi\varepsilon_0 r}\right) = \ell(\ell + 1)$$

We recognize the preceding equation as the Laguerre equation, and we recognize

$$\frac{1}{\Theta(\theta)}\frac{1}{\sin\theta}\frac{d}{d\theta}\left(\sin\theta\frac{d\Theta}{d\theta}\right) - \frac{m^2}{\sin^2\theta} = -\ell(\ell + 1)$$

as the associated Legendre equation.

SUGGESTED READINGS

1. BRADLEY, GERALD L., and SMITH, KARL J. *Calculus*. Upper Saddle River, NJ: Prentice Hall, 1995.
2. MARGENAU, HENRY, and MURPHY, GEORGE. *The Mathematics of Physics and Chemistry*. New York: D. Van Nostrand, 1943. This text is out of print; however, many libraries may still have a copy, and used copies sometimes can be found. If you can find a copy, grab it!
3. NAGLE, R. KENT, and SAFF, EDWARD B. *Fundamentals of Differential Equations and Boundary Value Problems*, 2d ed. Boston: Addison-Wesley, 1996.
4. VARBERG, DALE, and PURCELL, EDWIN J. *Calculus*, 7th ed. Upper Saddle River, NJ: Prentice Hall, 1997.

PROBLEMS

1. Solve the following linear differential equations:

(a) $\dfrac{dy}{dx} + 3y = 0$

(b) $\dfrac{dy}{dx} - 3y = 0$

(c) $\dfrac{d^2y}{dx^2} + 2\dfrac{dy}{dx} + y = 0$

(d) $\dfrac{d^2y}{dx^2} - 6\dfrac{dy}{dx} + 9y = 0$

(e) $\dfrac{d^2y}{dx^2} - 9y = 0$

(f) $\dfrac{d^2y}{dx^2} + 4y = 0$

(g) $\dfrac{dx}{dt} = k(a - x) - k_2x$; k_1, k_2, and a are constants.

(h) $\dfrac{d\phi}{dr} = -a\phi$; a is constant.

(i) $\dfrac{d(A)}{(A)} = -k\,dt$; k is constant.

(j) $\dfrac{1}{\Phi(\phi)}\dfrac{d^2\Phi}{d\phi^2} = -m^2$; m is constant.

(k) $m\dfrac{d^2y}{dt^2} = -ky$; m and k are constants.

(l) $\dfrac{d^2\psi}{dx^2} + \dfrac{8\pi^2 mE}{h^2}\psi = 0$; m, E, h, π are constants.

2. Test the following differentials for exactness:

(a) $dF = 2xy^2\,dx + 2yx^2\,dy$

(b) $dF = 8x\,dx$

(c) $dF = 12x^2y\,dx + 4x^3\,dy$

(d) $dF = 5\,dx$

(e) $dF = \dfrac{1}{y}dx - \dfrac{x}{y^2}dy$

(f) $dF = xy\,dx + x^3\,dy$

(g) $dP = \dfrac{nR}{V}\,dT - \dfrac{nRT}{V^2}\,dV$; n and R are constants.

(h) $dV = \pi r^2\,dh + 2\pi rh\,dr$

(i) $dq = nC_V\,dT + \dfrac{nRT}{v}\,dV$; n, C_V, and R are constants.

(j) $d\rho = -\dfrac{PM}{RT^2}\,dT + \dfrac{M}{RT}\,dP$; M and R are constants.

(k) $dE = nC_V\,dT + \dfrac{n^2a}{V^2}\,dV$; n, C_V, and a are constants.

3. Show that the inexact differential $dq = nC_V\,dT + (nRT/V)\,dV$, where n, C_V, and R are constants, can be made exact by multiplying by an integrating factor $1/T$. The resulting differential dS is called the *differential entropy change*.

4. Show that if sin $4x$ and cos $4x$ are particular solutions of the differential equation

$$\frac{d^2y}{dx^2} + 16y = 0$$

then a linear combination of the two solutions also is a solution of the equation.

5. Bessel's equation is an important differential equation having the general form

$$x^2\frac{d^2y}{dx^2} + x\frac{dy}{dx} + (x^2 - c^2)y = 0$$

where c is a constant. Find the indicial equation for the series solution of this equation.

6. A one-dimensional harmonic oscillator is described classically by the equation

$$\frac{d^2y}{dt^2} + 4\pi^2\nu^2y = 0$$

Show that $y = A \sin 2\pi\nu t$, where A, π, and ν are constants, is a solution of the one-dimensional harmonic oscillator.

7. The differential equation describing the spatial behavior of a one-dimensional wave is

$$\frac{d^2f}{dx^2} + \frac{4\pi^2}{\lambda^2}f(x) = 0$$

where λ is the wavelength. Find the general solution of this equation.

8. *Boundary conditions* are special restrictions imposed on the solutions of differential equations. The boundary conditions for a plucked string bound at both ends between $x = 0$ and $x = L$, and described by the equation given in Problem 7, are that $f(x) = 0$ at $x = 0$ and $x = L$. Show how the boundary conditions affect the solution of the equation in Problem 7.

9. Show that if we let $x = \cos\theta$, the solution of the associated Legendre's equation [Equation (7-47)] is $y = \cos\theta$ when $\ell = 1$ and $m = 0$ and is $y = \sin\theta$ when $\ell = 1$ and $m = 1$.

10. Show that if we let $x = \cos\theta$, the solution of the associated Legendre's equation [Equation (7-47)] is $y = \frac{1}{2}(3\cos^2\theta - 1)$, when $\ell = 2$ and $m = 0$.

11. The Schrödinger equation for a particle in a three-dimensional box is

$$\frac{\partial^2\psi}{\partial x^2} + \frac{\partial^2\psi}{\partial y^2} + \frac{\partial^2\psi}{\partial z^2} + \frac{8\pi^2mE}{h^2}\psi = 0$$

where E, m, and h are constants and $\psi = f(x, y, z)$. Separate the equation into an equation in x, an equation in y, and an equation in z, by assuming that

$$\psi(x, y, z) = f(x)g(y)h(z)$$

CHAPTER 8

Scalars and Vectors

8-1 INTRODUCTION

A *scalar* is a physical quantity that has *magnitude*, but no direction in space. A scalar can be specified by a single number. Examples of scalars are length, mass, temperature, and speed (not velocity). *Vectors*, by contrast, are quantities that have both *magnitude* and *direction*. Thus, velocity, which is a vector, has a magnitude, known as speed (10 m/s), and a direction (along the x-axis). That direction enters into a vector is easily verified when one considers that, to change from going, say, 20 miles per hour in a northerly direction to going 20 miles per hour in a westerly direction, a force (and, hence, an acceleration) must be applied to the wheels of a car. An acceleration is a change in velocity with respect to time. Thus, the velocity of the car changed, even though the speed of the car remained constant. While it is customary to represent a vector pictorially with an arrow, we find that, mathematically, it is more convenient to define a vector in terms of an ordered set of real numbers $\mathbf{A} = \langle a_x, a_y, a_z \rangle$, where a_x, a_y, and a_z are the components of the vector in Cartesian space.

8-2 ADDITION OF VECTORS

Since vectors are quantities having both magnitude and direction, the sum of two or more vectors (which is a vector) must have those properties as well. One method of adding vectors is the *triangle method*, shown in two dimensions in Fig. 8-1(a). In this method, the head of one vector **A** is placed against the tail of another vector **B**, forming a triangle. The sum of the vectors is a vector represented by an arrow from the tail of **A** to the head of **B**. Closely related to the triangle method is the *parallelogram method*, shown in two dimensions in Fig. 8-1(b). In this method, both vectors extend out from the origin of the coordinate system. A parallelogram is constructed, as shown in the diagram, and the sum of the two vectors is the length and direction of the diagonal of the parallelogram. Expressed mathematically, the sum of two vectors $\mathbf{A} = \langle a_x, a_y, a_z \rangle$ and $\mathbf{B} = \langle b_x, b_y, b_z \rangle$ is

$$\mathbf{A} + \mathbf{B} = \langle a_x + b_x, a_y + b_y, a_z + b_z \rangle \tag{8-1}$$

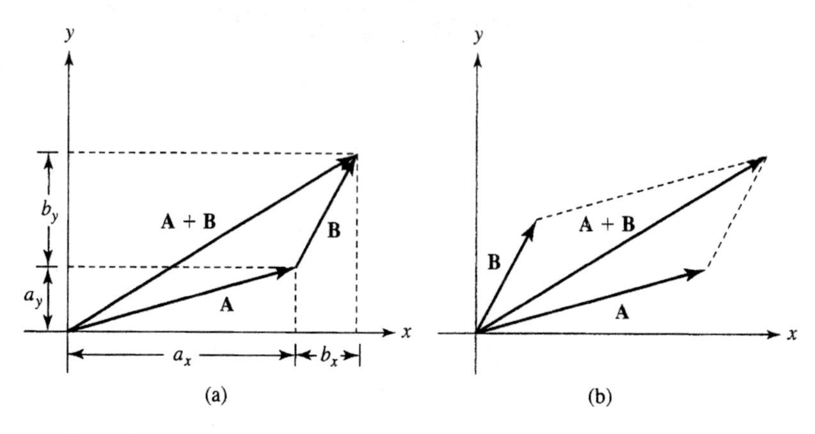

Figure 8-1 Addition of vectors by (a) triangle method. (b) parallelogram method.

Unit Vectors Once we know how to add vectors, we find it useful to represent a vector in terms of *unit vectors* lying along the coordinate axes of a coordinate system. (It is not necessary, however, that unit vectors lie along the coordinate axes in *all* coordinate systems.) Let us define **i**, **j**, and **k** as unit vectors—vectors having a magnitude (absolute value) of unity, lying along the x-, y-, and z-axes of a Cartesian coordinate system.[1] Using the method of addition, we find that any vector **A** can be described as the sum of multiples of these unit lengths **i**, **j**, and **k**. Thus, we can write

$$\mathbf{A} = \mathbf{i}\, a_x + \mathbf{j}\, a_y + \mathbf{k}\, a_z \tag{8-2}$$

where a_x, a_y, and a_z are the scalar multiples. In terms of the scalar multiples, the magnitude, or absolute value, of the vector and its direction can be found using the transformation equations for plane polar coordinates if the vector is confined to the xy-plane or the transformation equations for spherical polar coordinates if the vector lies in three dimensions. (See Chapter 1.) In three dimensions, we have

$$\begin{aligned} |A| &= (a_x^2 + a_y^2 + a_z^2)^{1/2} \\ \theta &= \cos^{-1}\left(\frac{a_z}{(a_x^2 + a_y^2 + a_z^2)^{1/2}}\right) \text{ from the } z\text{-axis} \\ \phi &= \tan^{-1}\left(\frac{a_y}{a_x}\right) \text{ from the } xz\text{-plane} \end{aligned} \tag{8-3}$$

[1]Some textbooks designate the unit vectors as $\hat{\imath}$, $\hat{\jmath}$, and $\hat{k}$ (read "*i* hat", "*j* hat", and "*k* hat").

Examples

(a) Determine the magnitude and direction of the vector $\mathbf{A}\langle 3, 4, 3\rangle = \mathbf{i}(3) + \mathbf{j}(4) + \mathbf{k}(3)$, illustrated in Fig. 8-2.

Solution

$$|A| = \sqrt{3^2 + 4^2 + 3^2} = \sqrt{34}$$

$$\theta = \cos^{-1}\frac{3}{\sqrt{34}} = 59.04° \text{ from } z\text{-axis}$$

$$\phi = \tan^{-1}\frac{4}{3} = 53.13° \text{ from } x\text{-}z \text{ plane}$$

(b) Determine the sum $\mathbf{C} = \mathbf{A} + \mathbf{B}$, where $\mathbf{A} = \mathbf{i}(2) + \mathbf{j}(3) + \mathbf{k}(5)$ and $\mathbf{B} = \mathbf{i}(4) + \mathbf{j}(-2) + \mathbf{k}(-1)$.

Solution

$$\mathbf{C} = \mathbf{A} + \mathbf{B} = \mathbf{i}(a_x + b_x) + \mathbf{j}(a_y + b_y) + \mathbf{k}(a_z + b_z)$$
$$= \mathbf{i}(2 + 4) + \mathbf{j}(3 - 2) + \mathbf{k}(5 - 1) = \mathbf{i}(6) + \mathbf{j}(1) + \mathbf{k}(4)$$

$$|C| = \sqrt{6^2 + 1^2 + 4^2} = \sqrt{53} = 7.28$$

$$\theta = \cos^{-1}\frac{4}{7.28} = 56.67° \text{ from } z\text{-axis}$$

$$\phi = \tan^{-1}\frac{1}{6} = 9.46° \text{ from } xz\text{-plane}$$

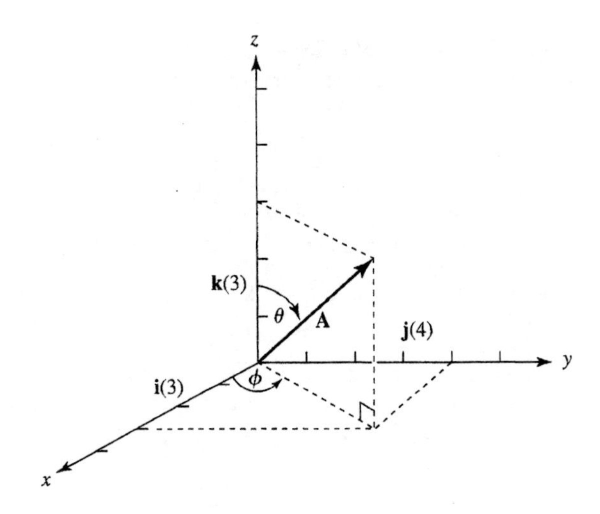

Figure 8-2 Vector $\mathbf{A}\langle 3, 4, 3\rangle = \mathbf{i}(3) + \mathbf{j}(4) + \mathbf{k}(3)$.

8-3 MULTIPLICATION OF VECTORS

There are at least two ways to multiply vectors. The first way is to find the scalar, or "dot," product between the two vectors. The second way is to find the vector, or "cross," product between the two vectors.

Scalar Multiplication The scalar, or "dot," product is defined by the equation

$$\mathbf{A}\cdot\mathbf{B} = |A||B| \cos\theta_{AB} \tag{8-4}$$

where θ_{AB} is the angle between **A** and **B**. This product (read "*A* dot *B*") is called the scalar product because the result of the multiplication yields a scalar. To see this, consider first the scalar products between the unit vectors **i**, **j**, and **k**. Since the angle between any two unit vectors in Cartesian coordinates is 90°, and the magnitude of the vectors is unity, we can write

$$\begin{aligned} \mathbf{i}\cdot\mathbf{i} &= 1 \qquad & \mathbf{j}\cdot\mathbf{j} &= 1 \\ \mathbf{i}\cdot\mathbf{j} &= 0 \qquad & \mathbf{j}\cdot\mathbf{k} &= 0 \\ \mathbf{i}\cdot\mathbf{k} &= 0 \qquad & \mathbf{k}\cdot\mathbf{k} &= 1 \end{aligned} \tag{8-5}$$

Vectors obeying Equations (8-5) are said to be *orthogonal*. In general, we can state that if two vectors $\mathbf{q}_i$ and $\mathbf{q}_j$ are orthogonal unit vectors, then

$$\mathbf{q}_i\cdot\mathbf{q}_j = \delta_{ij}, \quad \text{where } \delta_{ij} = \begin{cases} 1 \text{ for } i = j \\ 0 \text{ for } i \neq j \end{cases} \tag{8-6}$$

We recognize δ_{ij} as the *Kronecker delta*. (See Chapter 6.)

Now consider the scalar product between two vectors $\mathbf{A} = \mathbf{i}\,a_x + \mathbf{j}\,a_y + \mathbf{k}\,a_z$ and $\mathbf{B} = \mathbf{i}\,b_x + \mathbf{j}\,b_y + \mathbf{k}\,b_z$:

$$\begin{aligned} \mathbf{A}\cdot\mathbf{B} &= (\mathbf{i}\,a_x + \mathbf{j}\,a_y + \mathbf{k}\,a_z)\cdot(\mathbf{i}\,b_x + \mathbf{j}\,b_y + \mathbf{k}\,b_z) \\ &= a_xb_x\,\mathbf{i}\cdot\mathbf{i} + a_xb_y\,\mathbf{i}\cdot\mathbf{j} + a_xb_z\,\mathbf{i}\cdot\mathbf{k} + a_yb_x\,\mathbf{j}\cdot\mathbf{i} + a_yb_y\,\mathbf{j}\cdot\mathbf{j} + a_yb_z\,\mathbf{j}\cdot\mathbf{k} \\ &\quad + a_zb_x\,\mathbf{k}\cdot\mathbf{i} + a_zb_y\,\mathbf{k}\cdot\mathbf{j} + a_zb_z\,\mathbf{k}\cdot\mathbf{k} \end{aligned}$$

Taking Equations (8-5) into account, we have

$$\mathbf{A}\cdot\mathbf{B} = a_xb_x + a_yb_y + a_zb_z \tag{8-7}$$

which, indeed, is a scalar.

Example

Find the scalar product between the vectors $\mathbf{A}\langle 1, 3, 2\rangle$ and $\mathbf{B}\langle 4, -4, 1\rangle$.

Solution

$$\mathbf{A}\cdot\mathbf{B} = (1)(4) + (3)(-4) + (2)(1) = -6$$

Vector Multiplication The vector, or "cross," product is defined by the equation

$$\mathbf{A} \times \mathbf{B} = |A||B|\,\mathbf{C}\sin\theta_{AB} \qquad (8\text{-}8)$$

where θ_{AB} is the angle between **A** and **B**, and **C** is a unit vector perpendicular to the plane formed by **A** and **B**. This product is called the vector product because the result of the multiplication is a vector. To obtain the direction of $\mathbf{A} \times \mathbf{B}$ (read "*A* cross *B*"), the "right-hand rule" can be used. The rule states that if the fingers of the right hand rotate **A** into **B** through the smaller angle between their positive senses, the thumb will point in the direction of the cross product, shown in Fig. 8-3.

With the preceding discussion in mind, let us determine the vector products between the unit vectors **i, j,** and **k**:

$$\begin{array}{lll} \mathbf{i}\times\mathbf{i} = 0 & \mathbf{i}\times\mathbf{j} = \mathbf{k} & \mathbf{j}\times\mathbf{j} = 0 \\ \mathbf{j}\times\mathbf{k} = \mathbf{i} & \mathbf{k}\times\mathbf{k} = 0 & \mathbf{k}\times\mathbf{i} = \mathbf{j} \end{array} \qquad (8\text{-}9)$$

From these equations, we can obtain an expression for the vector product between vectors **A** and **B**:

$$\begin{aligned} \mathbf{A} \times \mathbf{B} &= (\mathbf{i}\,a_x + \mathbf{j}\,a_y + \mathbf{k}\,a_z) \times (\mathbf{i}\,b_x + \mathbf{j}\,b_y + \mathbf{k}\,b_z) \\ &= a_xb_x\,\mathbf{i}\times\mathbf{i} + a_xb_y\,\mathbf{i}\times\mathbf{j} + a_xb_z\,\mathbf{i}\times\mathbf{k} + a_yb_x\,\mathbf{j}\times\mathbf{i} + a_yb_y\,\mathbf{j}\times\mathbf{j} \\ &\quad + a_yb_z\,\mathbf{j}\times\mathbf{k} + a_zb_x\,\mathbf{k}\times\mathbf{i} + a_zb_y\,\mathbf{k}\times\mathbf{j} + a_zb_z\,\mathbf{k}\times\mathbf{k} \end{aligned}$$

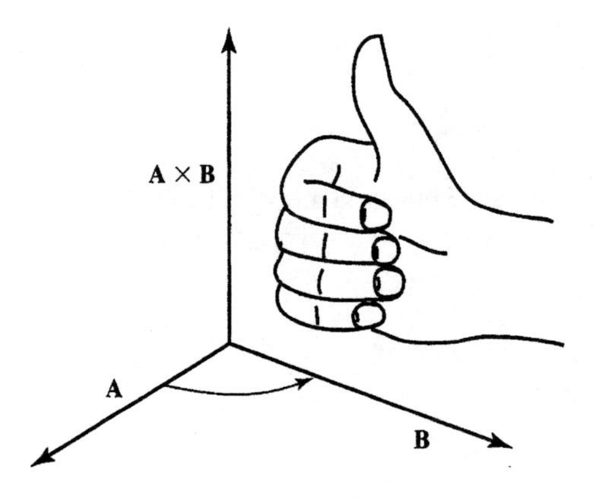

Figure 8-3 Direction of vector product **A** × **B** using the right hand rule.

Taking into account Equation (8-9), we have

$$\mathbf{A} \times \mathbf{B} = \mathbf{i}(a_y b_z - a_z b_y) + \mathbf{j}(a_z b_x - a_x b_z) + \mathbf{k}(a_x b_y - a_y b_x) \quad (8\text{-}10)$$

A convenient way to remember $\mathbf{A} \times \mathbf{B}$ is to express it in the form of a determinant (see Chapter 9):

$$\mathbf{A} \times \mathbf{B} = \begin{vmatrix} \mathbf{i} & \mathbf{j} & \mathbf{k} \\ a_x & a_y & a_z \\ b_x & b_y & b_z \end{vmatrix}$$

Example

Determine the vector product between the vectors $\mathbf{A}\langle 4, 3, 2\rangle$ and $\mathbf{B}\langle -1, 2, -3\rangle$.

Solution

$$\mathbf{A} \times \mathbf{B} = \begin{vmatrix} \mathbf{i} & \mathbf{j} & \mathbf{k} \\ 4 & 3 & 2 \\ -1 & 2 & -3 \end{vmatrix}$$

$$\begin{aligned} \mathbf{A} \times \mathbf{B} &= \mathbf{i}\,[(3)(-3) - (2)(2)] - \mathbf{j}\,[(4)(-3) \\ &\quad - (-1)(2)] + \mathbf{k}\,[(4)(2) - (-1)(3)] \\ &= \mathbf{i}\,(-9 - 4) - \mathbf{j}\,(-12 + 2) + \mathbf{k}\,(8 + 3) \\ &= \mathbf{i}\,(-13) + \mathbf{j}\,(10) + \mathbf{k}\,(11) \end{aligned}$$

One very important algebraic property of the vector product is that it is not commutative. Stated mathematically, $\mathbf{A} \times \mathbf{B} \neq \mathbf{B} \times \mathbf{A}$. It is not difficult to show that, in fact, vector multiplication is *anti*commutative: $\mathbf{A} \times \mathbf{B} = -\mathbf{B} \times \mathbf{A}$. Therefore, when performing vector multiplication, one must be careful to preserve the order of the factors.

8-4 APPLICATIONS

In this section, we shall consider examples that demonstrate the application of vector analysis to physicochemical systems. The subject of vector operators is discussed in Chapter 10.

1. The interaction between the magnetic moment of a nucleus, $\boldsymbol{\mu}$, and a magnetic field $\mathbf{H}$ is given by the equation

$$E = -\boldsymbol{\mu} \cdot \mathbf{H}$$

Let us determine the components of energy associated with the interaction along the x-, y-, and z-axes. Since

$$\boldsymbol{\mu} = \mathbf{i}\,\mu_x + \mathbf{j}\,\mu_y + \mathbf{k}\,\mu_z$$
$$\mathbf{H} = \mathbf{i}\,H_x + \mathbf{j}\,H_y + \mathbf{k}\,H_z$$

it follows that

$$\boldsymbol{\mu}\cdot\mathbf{H} = \mu_x H_x + \mu_y H_y + \mu_z H_z$$

Also, since

$$E = E_x + E_y + E_z$$

we can write

$$E_x = -\mu_x H_x,\ E_y = -\mu_y H_y, \text{ and } E_z = -\mu_z H_z$$

Notice that although both $\boldsymbol{\mu}$ and $\mathbf{H}$ are vectors, energy is a scalar.

2. The torque exerted on a nucleus having a magnetic moment $\boldsymbol{\mu}$ in a magnetic field $\mathbf{H}$ is

$$\mathbf{T} = -\boldsymbol{\mu}\times\mathbf{H}$$

Torque, however, is the rate of change of angular momentum, $d\mathbf{L}/dt$, where

$$\mathbf{L} = \mathbf{i}\,L_x + \mathbf{j}\,L_y + \mathbf{k}\,L_z$$

and

$$\frac{d\mathbf{L}}{dt} = \mathbf{i}\frac{dL_x}{dt} + \mathbf{j}\frac{dL_y}{dt} + \mathbf{k}\frac{dL_z}{dt}$$

Let us determine the torque on the nucleus along the x-, y-, and z-axes. We have

$$\frac{d\mathbf{L}}{dt} = -\boldsymbol{\mu}\times\mathbf{H}$$

$$\boldsymbol{\mu}\times\mathbf{H} = \begin{vmatrix} \mathbf{i} & \mathbf{j} & \mathbf{k} \\ \mu_x & \mu_y & \mu_z \\ H_x & H_y & H_z \end{vmatrix}$$

$$\begin{aligned}\boldsymbol{\mu}\times\mathbf{H} &= \mathbf{i}\,[\mu_y H_z - \mu_z H_y] - \mathbf{j}\,[\mu_x H_z - \mu_z H_x] + \mathbf{k}\,[\mu_x H_y - \mu_y H_x] \\ &= \mathbf{i}\,[\mu_y H_z - \mu_z H_y] + \mathbf{j}\,[\mu_z H_x - \mu_x H_z] + \mathbf{k}\,[\mu_x H_y - \mu_y H_x] \\ -\boldsymbol{\mu}\times\mathbf{H} &= \mathbf{i}\,[\mu_z H_y - \mu_y H_z] + \mathbf{j}\,[\mu_x H_z - \mu_z H_x] + \mathbf{k}\,[\mu_y H_x - \mu_x H_y]\end{aligned}$$

$$\frac{dL_x}{dt} = \mu_z H_y - \mu_y H_z,\ \frac{dL_y}{dt} = \mu_x H_z - \mu_z H_x, \text{ and } \frac{dL_z}{dt} = \mu_y H_x - \mu_x H_y$$

SUGGESTED READINGS

1. BRADLEY, GERALD L., and SMITH, KARL J. *Calculus*. Upper Saddle River, NJ: Prentice Hall, 1995.
2. VARBERG, DALE, and PURCELL, EDWIN J. *Calculus*, 7th ed., Upper Saddle River, NJ: Prentice Hall, 1997.
3. WASHINGTON, ALLYN J. *Basic Technical Mathematics*, 6th ed. Boston: Addison-Wesley, 1995.

PROBLEMS

1. Determine the magnitude and direction of the following vectors:
   (a) $\mathbf{A}\langle 1, 3\rangle$
   (b) $\mathbf{A}\langle 2, 2\rangle$
   (c) $\mathbf{A}\langle 3, -4\rangle$
   (d) $\mathbf{A}\langle -2, 0\rangle$
   (e) $\mathbf{A}\langle -1, -6\rangle$
   (f) $\mathbf{A}\langle 1, 1, 3\rangle$
   (g) $\mathbf{A}\langle 2, 3, 4\rangle$
   (h) $\mathbf{A}\langle -1, 2, -1\rangle$
   (i) $\mathbf{A}\langle -1, -1, -3\rangle$
   (j) $\mathbf{A}\langle 1, 0, -1\rangle$
2. For each of the following sums, determine the magnitude and direction of the resultant vector:
   (a) $\mathbf{A}\langle 1, 3\rangle + \mathbf{B}\langle 3, 1\rangle$
   (b) $\mathbf{A}\langle -1, 2\rangle + \mathbf{B}\langle 2, 2\rangle$
   (c) $\mathbf{A}\langle 3, -1\rangle + \mathbf{B}\langle 0, 4\rangle$
   (d) $\mathbf{A}\langle 1, 1, 1\rangle + \mathbf{B}\langle 2, 3, 4\rangle$
   (e) $\mathbf{A}\langle -2, 3, 4\rangle + \mathbf{B}\langle -1, -4, -6\rangle$
   (f) $\mathbf{A}\langle 2, 0, 3\rangle + \mathbf{B}\langle -3, 6, -9\rangle$
3. Find the following scalar products:
   (a) $\mathbf{A}\langle 1, 3\rangle \cdot \mathbf{B}\langle 3, 1\rangle$
   (b) $\mathbf{A}\langle -1, 2\rangle \cdot \mathbf{B}\langle 2, 2\rangle$
   (c) $\mathbf{A}\langle 3, -1\rangle \cdot \mathbf{B}\langle 0, 4\rangle$
   (d) $\mathbf{A}\langle 1, 1, 1\rangle \cdot \mathbf{B}\langle 2, 3, 4\rangle$
   (e) $\mathbf{A}\langle -2, 3, 4\rangle \cdot \mathbf{B}\langle -1, -4, -6\rangle$
   (f) $\mathbf{A}\langle 2, 0, 3\rangle \cdot \mathbf{B}\langle -3, 6, -9\rangle$
4. Find the magnitude and direction of the following vector products:
   (a) $\mathbf{A}\langle 1, 3\rangle \times \mathbf{B}\langle 3, 1\rangle$
   (b) $\mathbf{A}\langle -1, 2\rangle \times \mathbf{B}\langle 2, 2\rangle$
   (c) $\mathbf{A}\langle 3, -1\rangle \times \mathbf{B}\langle 0, 4\rangle$
   (d) $\mathbf{A}\langle 1, 1, 1\rangle \times \mathbf{B}\langle 2, 3, 4\rangle$
   (e) $\mathbf{A}\langle -2, 3, 4\rangle \times \mathbf{B}\langle -1, -4, -6\rangle$
   (f) $\mathbf{A}\langle 2, 0, 3\rangle \times \mathbf{B}\langle -3, 6, -9\rangle$
5. Show that

$$\mathbf{A} + (\mathbf{B} + \mathbf{C}) = (\mathbf{A} + \mathbf{B}) + \mathbf{C}$$

6. Show that scalar multiplication is commutative and vector multiplication is not. That is, show that

$$\mathbf{A} \cdot \mathbf{B} = \mathbf{B} \cdot \mathbf{A} \quad \text{but} \quad \mathbf{A} \times \mathbf{B} \neq \mathbf{B} \times \mathbf{A}$$

7. Show that

$$\mathbf{A} \cdot \mathbf{A} = |A|^2$$

8. Angular momentum is given by the equation $\mathbf{L} = \mathbf{r} \times \mathbf{p}$, where $\mathbf{r} = \mathbf{i}\,x + \mathbf{j}\,y + \mathbf{k}\,z$ is the radius of curvature of the motion and $\mathbf{p} = \mathbf{i}\,p_x + \mathbf{j}\,p_y + \mathbf{k}\,p_z$ is the linear momentum of the body in motion. Assuming that

$$\mathbf{L} = \mathbf{i}\,L_x + \mathbf{j}\,L_y + \mathbf{k}\,L_z$$

find the components of angular momentum in the x, y, and z directions.

9. Show that the vectors $\mathbf{A} = \frac{1}{2}\mathbf{q}_1 + \frac{1}{2}\mathbf{q}_2 + \frac{1}{2}\mathbf{q}_3 + \frac{1}{2}\mathbf{q}_4$ and $\mathbf{B} = \frac{1}{2}\mathbf{q}_1 - \frac{1}{2}\mathbf{q}_2 + \frac{1}{2}\mathbf{q}_3 - \frac{1}{2}\mathbf{q}_4$, where $\mathbf{q}_1$, $\mathbf{q}_2$, $\mathbf{q}_3$, and $\mathbf{q}_4$ are unit vectors, are orthogonal.

CHAPTER 9

Matrices and Determinants

9-1 INTRODUCTION

In certain areas of physical chemistry, it is convenient to utilize a two-dimensional array of numbers called a *matrix*. Matrices may be either square, containing an equal number of horizontal and vertical lines, or rectangular. The horizontal lines are called *rows*, the vertical lines *columns*. A matrix with m rows and n columns is represented by the expression

$$\mathbf{A} = \begin{pmatrix} a_{11} & a_{12} & a_{13} & \cdots & a_{1n} \\ a_{21} & a_{22} & a_{23} & \cdots & a_{2n} \\ \vdots & & & & \vdots \\ a_{m1} & a_{m2} & a_{m3} & \cdots & a_{mn} \end{pmatrix}$$

where $a_{ij} = a_{11}, a_{12}, a_{13}, \ldots$ are known as the *elements* of the matrix. Such a matrix is called either a *matrix of order (m,n)* or an *$m \times n$ matrix*. The simplest forms of matrices are the row matrix

$$\mathbf{B} = (b_1\ b_2\ b_3\ \ldots\ b_n)$$

and the column matrix

$$\mathbf{C} = \begin{pmatrix} c_1 \\ c_2 \\ \vdots \\ c_n \end{pmatrix}$$

Matrices have some highly useful properties and an algebra all their own. However, before going into these topics, let us concentrate on one specific type of matrix: the square matrix.

9-2 SQUARE MATRICES AND DETERMINANTS

As mentioned, a square matrix is a matrix with an equal number of rows and columns. The number of rows or columns in a square matrix is called the *order* of the matrix. Hence, a third-order matrix is a matrix with three rows and three columns.

Associated with every square matrix is a real number called the *determinant* of the matrix. The determinant of a second-order matrix is

$$D = \begin{vmatrix} a & b \\ c & d \end{vmatrix} = ad - bc \tag{9-1}$$

where a, b, c, and d are the elements of the matrix. It is important to note that the matrix itself has no numerical value; only the determinant of the matrix can be assigned a specific value. To illustrate the use of Equation (9-1), consider the following examples:

Examples

(a) Evaluate the determinant $\begin{vmatrix} 3 & 1 \\ 4 & 2 \end{vmatrix}$.

Solution

$$\begin{vmatrix} 3 & 1 \\ 4 & 2 \end{vmatrix} = (3)(2) - (4)(1) = 2$$

(b) Evaluate the determinant $\begin{vmatrix} 5 & 6 \\ -1 & -4 \end{vmatrix}$.

Solution

$$\begin{vmatrix} 5 & 6 \\ -1 & -4 \end{vmatrix} = (5)(-4) - (-1)(6) = -14$$

(c) Evaluate the determinant $\begin{vmatrix} \sin\theta & \cos\theta \\ -\cos\theta & \sin\theta \end{vmatrix}$.

Solution

$$\begin{vmatrix} \sin\theta & \cos\theta \\ -\cos\theta & \sin\theta \end{vmatrix} = \sin^2\theta + \cos^2\theta = 1$$

To evaluate determinants of orders higher than 2, we use the *method of cofactors*. A special shortcut method for solving 3×3 determinants is given in Appendix V.

Consider the determinant

$$D = \begin{vmatrix} a_{11} & a_{12} & a_{13} \\ a_{21} & a_{22} & a_{23} \\ a_{31} & a_{32} & a_{33} \end{vmatrix}$$

The cofactor of element a_{ij} is equal to $(-1)^{i+j}$, multiplied by the determinant that is formed by eliminating the *ith* row and the *jth* column from the original determinant. The determinant is then expanded by summing together the elements, multiplied by their cofactors, for any row. Hence, using the first row, for example, we can write

$$D = \begin{vmatrix} a_{11} & a_{12} & a_{13} \\ a_{21} & a_{22} & a_{23} \\ a_{31} & a_{32} & a_{33} \end{vmatrix} = a_{11}\begin{vmatrix} a_{22} & a_{23} \\ a_{32} & a_{33} \end{vmatrix} - a_{12}\begin{vmatrix} a_{21} & a_{23} \\ a_{31} & a_{33} \end{vmatrix} + a_{13}\begin{vmatrix} a_{21} & a_{22} \\ a_{31} & a_{32} \end{vmatrix}$$

Similar expressions follow when the elements of the second row or third row are used as cofactors.

If each new determinant that results from an expansion by cofactors is of order higher than 2, then the process is repeated on each such determinant until all the resulting determinants are of order 2. Since, in this example, the resulting determinants are already of order 2, we have

$$D = a_{11}(a_{22}a_{33} - a_{32}a_{23}) - a_{12}(a_{21}a_{33} - a_{31}a_{23}) + a_{13}(a_{21}a_{32} - a_{31}a_{22})$$

Example

Evaluate the determinant $\begin{vmatrix} 1 & 4 & 3 & 2 \\ 6 & 1 & 1 & 3 \\ -1 & 4 & 5 & -6 \\ 2 & 1 & 2 & -3 \end{vmatrix}$.

Solution. Expanding by cofactors, we obtain

$$\begin{vmatrix} 1 & 4 & 3 & 2 \\ 6 & 1 & 1 & 3 \\ -1 & 4 & 5 & -6 \\ 2 & 1 & 2 & -3 \end{vmatrix} = (1)\begin{vmatrix} 1 & 1 & 3 \\ 4 & 5 & -6 \\ 1 & 2 & -3 \end{vmatrix} - (4)\begin{vmatrix} 6 & 1 & 3 \\ -1 & 5 & -6 \\ 2 & 2 & -3 \end{vmatrix}$$

$$+ (3)\begin{vmatrix} 6 & 1 & 3 \\ -1 & 4 & -6 \\ 2 & 1 & -3 \end{vmatrix} - (2)\begin{vmatrix} 6 & 1 & 1 \\ -1 & 4 & 5 \\ 2 & 1 & 2 \end{vmatrix}$$

$$= (1)(1)\begin{vmatrix} 5 & -6 \\ 2 & -3 \end{vmatrix} - (1)(1)\begin{vmatrix} 4 & -6 \\ 1 & -3 \end{vmatrix}$$

$$+ (1)(3)\begin{vmatrix} 4 & 5 \\ 1 & 2 \end{vmatrix} - (4)(6)\begin{vmatrix} 5 & -6 \\ 2 & -3 \end{vmatrix}$$

$$+ (4)(1)\begin{vmatrix} -1 & -6 \\ 2 & -3 \end{vmatrix} - (4)(3)\begin{vmatrix} -1 & 5 \\ 2 & 2 \end{vmatrix}$$

$$+ (3)(6)\begin{vmatrix} 4 & -6 \\ 1 & -3 \end{vmatrix} - (3)(1)\begin{vmatrix} -1 & -6 \\ 2 & -3 \end{vmatrix}$$

$$+ (3)(3)\begin{vmatrix} -1 & 4 \\ 2 & 1 \end{vmatrix} - (2)(6)\begin{vmatrix} 4 & 5 \\ 1 & 2 \end{vmatrix}$$

$$+ (2)(1)\begin{vmatrix} -1 & 5 \\ 2 & 2 \end{vmatrix} - (2)(1)\begin{vmatrix} -1 & 4 \\ 2 & 1 \end{vmatrix}$$

$$= (1)(-3) - (1)(-6) + (3)(3) - (24)(-3)$$
$$+ (4)(15) - (12)(-12) + (18)(-6)$$
$$- (3)(15) + (9)(-9) - (12)(3)$$
$$+ (2)(-12) - (2)(-9) = 12$$

9-3 MATRIX ALGEBRA

Let us turn now to several rules that govern the properties of matrices and their determinants. These rules are presented without proof.

1. *If the corresponding rows and columns of a square matrix are interchanged, the determinant of the matrix remains unchanged:*

$$\begin{vmatrix} a_{11} & a_{12} & a_{13} \\ a_{21} & a_{22} & a_{23} \\ a_{31} & a_{32} & a_{33} \end{vmatrix} = \begin{vmatrix} a_{11} & a_{21} & a_{31} \\ a_{12} & a_{22} & a_{32} \\ a_{13} & a_{23} & a_{33} \end{vmatrix}$$

2. *If any two rows or columns of a determinant are interchanged, the sign of the determinant changes:*

$$\begin{vmatrix} a_{11} & a_{12} & a_{13} \\ a_{21} & a_{22} & a_{23} \\ a_{31} & a_{32} & a_{33} \end{vmatrix} = -\begin{vmatrix} a_{11} & a_{13} & a_{12} \\ a_{21} & a_{23} & a_{22} \\ a_{31} & a_{33} & a_{32} \end{vmatrix}$$

3. *If any two rows or columns of a square matrix are identical, its determinant is zero:*

$$\begin{vmatrix} a_{11} & a_{11} & a_{13} \\ a_{21} & a_{21} & a_{23} \\ a_{31} & a_{31} & a_{33} \end{vmatrix} = 0$$

4. *If each element in any row or column in a determinant is multiplied by the same number k, the value of the determinant is multiplied by k:*

$$\begin{vmatrix} a_{11} & a_{12} & a_{13} \\ ka_{21} & ka_{22} & ka_{23} \\ a_{31} & a_{32} & a_{33} \end{vmatrix} = k \begin{vmatrix} a_{11} & a_{12} & a_{13} \\ a_{21} & a_{22} & a_{23} \\ a_{31} & a_{32} & a_{33} \end{vmatrix}$$

5. *If each element in any row or column in a determinant is multiplied by the same number k, and the product is added to a corresponding element in another column, the value of the determinant remains unchanged:*

$$\begin{vmatrix} a_{11} + ka_{12} & a_{12} & a_{13} \\ a_{21} + ka_{22} & a_{22} & a_{23} \\ a_{31} + ka_{32} & a_{32} & a_{33} \end{vmatrix} = \begin{vmatrix} a_{11} & a_{12} & a_{13} \\ a_{21} & a_{22} & a_{23} \\ a_{31} & a_{32} & a_{33} \end{vmatrix}$$

6. *Two matrices are added by the addition of their elements. Matrix addition is defined only if the number of rows and columns of the first matrix is the same as the number of rows and columns of the second matrix. Thus,*

$$\begin{pmatrix} a_{11} & a_{12} & a_{13} \\ a_{21} & a_{22} & a_{23} \end{pmatrix} + \begin{pmatrix} b_{11} & b_{12} & b_{13} \\ b_{21} & b_{22} & b_{23} \end{pmatrix} = \begin{pmatrix} a_{11} + b_{11} & a_{12} + b_{12} & a_{13} + b_{13} \\ a_{21} + b_{21} & a_{22} + b_{22} & a_{23} + b_{23} \end{pmatrix}$$

7. *Two matrices are multiplied according to the following method:*

$$\begin{pmatrix} a_{11} & a_{12} & a_{13} \\ a_{21} & a_{22} & a_{23} \\ a_{31} & a_{32} & a_{33} \end{pmatrix} \begin{pmatrix} b_{11} & b_{12} & b_{13} \\ b_{21} & b_{22} & b_{23} \\ b_{31} & b_{32} & b_{33} \end{pmatrix} =$$

$$\begin{pmatrix} a_{11}b_{11} + a_{12}b_{21} + a_{13}b_{31} & a_{11}b_{12} + a_{12}b_{22} + a_{13}b_{32} & a_{11}b_{13} + a_{12}b_{23} + a_{13}b_{33} \\ a_{21}b_{11} + a_{22}b_{21} + a_{23}b_{31} & a_{21}b_{12} + a_{22}b_{22} + a_{23}b_{32} & a_{21}b_{13} + a_{22}b_{23} + a_{23}b_{33} \\ a_{31}b_{11} + a_{32}b_{21} + a_{33}b_{31} & a_{31}b_{12} + a_{32}b_{22} + a_{33}b_{32} & a_{31}b_{13} + a_{32}b_{23} + a_{33}b_{33} \end{pmatrix}$$

We find from experience that matrix multiplication is defined only if the number of columns of the first matrix equals the number of rows of the second matrix.

Moreover, if **A** is an $m \times r$ matrix with elements a_{ij} and **B** is an $r \times n$ matrix with elements b_{ij}, then the resulting product $\mathbf{C} = \mathbf{AB}$ will be an $m \times n$ matrix with elements

$$c_{ij} = \sum_k a_{ik} b_{kj}$$

It is important to note that, of the seven rules just given, this seventh is the most basic matrix property, since, as a natural consequence of the method of multiplication it expresses, it turns out that matrix multiplication is not necessarily commutative. That is, **AB** does not necessarily equal **BA**. This basic property of matrices ultimately led Werner Heisenberg and Max Born to what we now refer to as the Heisenberg uncertainty principle.

Examples

(a) $$\begin{pmatrix} 1 & 2 \\ 3 & 4 \end{pmatrix}\begin{pmatrix} 5 & 6 \\ 7 & 8 \end{pmatrix} = \begin{pmatrix} (1)(5) + (2)(7) & (1)(6) + (2)(8) \\ (3)(5) + (4)(7) & (3)(6) + (4)(8) \end{pmatrix} = \begin{pmatrix} 19 & 22 \\ 43 & 50 \end{pmatrix}$$

(b) $$\begin{pmatrix} a_{11} & a_{12} \\ a_{21} & a_{22} \end{pmatrix}\begin{pmatrix} c_1 \\ c_2 \end{pmatrix} = \begin{pmatrix} a_{11}c_1 + a_{12}c_2 \\ a_{21}c_1 + a_{22}c_2 \end{pmatrix}$$

(c) $$(a_1 \quad a_2)\begin{pmatrix} c_1 \\ c_2 \end{pmatrix} = (a_1c_1 + a_2c_2)$$

A special type of matrix multiplication occurs when matrices have nonzero elements in blocks along the diagonal. For example, consider the multiplication of the matrices

$$\left(\begin{array}{cc|c|cc} 1 & 2 & 0 & 0 & 0 \\ 3 & 4 & 0 & 0 & 0 \\ \hline 0 & 0 & 8 & 0 & 0 \\ \hline 0 & 0 & 0 & 1 & 4 \\ 0 & 0 & 0 & -5 & 6 \end{array}\right)\left(\begin{array}{cc|c|cc} -3 & -3 & 0 & 0 & 0 \\ 4 & 0 & 0 & 0 & 0 \\ \hline 0 & 0 & 2 & 0 & 0 \\ \hline 0 & 0 & 0 & 4 & -1 \\ 0 & 0 & 0 & -1 & 3 \end{array}\right)$$

In this special case, each set of blocks can be multiplied separately:

$$\begin{pmatrix} 1 & 2 \\ 3 & 4 \end{pmatrix}\begin{pmatrix} -3 & -3 \\ 4 & 0 \end{pmatrix} = \begin{pmatrix} 5 & -3 \\ 7 & -9 \end{pmatrix}$$

$$(8)(2) = (16)$$

$$\begin{pmatrix} 1 & 4 \\ -5 & 6 \end{pmatrix}\begin{pmatrix} 4 & -1 \\ -1 & 3 \end{pmatrix} = \begin{pmatrix} 0 & 11 \\ -26 & 23 \end{pmatrix}$$

The answer obtained is

$$\begin{pmatrix} 5 & -3 & 0 & 0 & 0 \\ 7 & -9 & 0 & 0 & 0 \\ 0 & 0 & 16 & 0 & 0 \\ 0 & 0 & 0 & 0 & 11 \\ 0 & 0 & 0 & -26 & 23 \end{pmatrix}$$

A matrix with nonzero elements along its diagonal and zero for all other elements is known as a *diagonal matrix*. When the nonzero elements along the diagonal are equal to unity, the matrix is called a *unit matrix*. Hence, a 3×3 unit matrix is

$$\mathbf{1} = \begin{pmatrix} 1 & 0 & 0 \\ 0 & 1 & 0 \\ 0 & 0 & 1 \end{pmatrix}$$

We can show by matrix multiplication that if $\mathbf{A}$ is any matrix, then

$$\mathbf{1\,A} = \mathbf{A}$$

If $\mathbf{A}$ is a square matrix and the determinant of $\mathbf{A}$ does not equal zero, then $\mathbf{A}$ is said to be *nonsingular*.[1] If $\mathbf{A}$ is a nonsingular matrix, then there exists an inverse matrix $\mathbf{A}^{-1}$ such that

$$\mathbf{A}^{-1}\mathbf{A} = \mathbf{A\,A}^{-1} = \mathbf{1}$$

The inverse of a matrix can be found by solving the equation $\mathbf{A}^{-1}\mathbf{A} = \mathbf{1}$, which is most conveniently done with the help of a computer program.

Example

Determine the inverse of the nonsingular matrix $\begin{pmatrix} 0 & -1 \\ 1 & 0 \end{pmatrix}$.

Solution. We must solve the equation

$$\begin{pmatrix} a & b \\ c & d \end{pmatrix}\begin{pmatrix} 0 & -1 \\ 1 & 0 \end{pmatrix} = \begin{pmatrix} 1 & 0 \\ 0 & 1 \end{pmatrix}$$

Using the rules of matrix multiplication, we obtain

$$0 + b = 1;\ \text{therefore, } b = 1$$
$$-a + 0 = 0;\ \text{therefore, } a = 0$$
$$0 + d = 0;\ \text{therefore, } d = 0$$
$$-c + 0 = 1;\ \text{therefore, } c = -1$$

[1]A singular matrix is defined as a matrix whose determinant is equal to zero.

Thus, the inverse of the given matrix is $\begin{pmatrix} 0 & 1 \\ -1 & 0 \end{pmatrix}$.

9-4 SOLUTIONS OF SYSTEMS OF LINEAR EQUATIONS

Consider the following set of linear equations:

$$\begin{array}{ccccccccccc} a_{11}x_1 & + & a_{12}x_2 & + & a_{13}x_3 & + & \cdots & + & a_{1n}x_n & = & c_1 \\ a_{21}x_1 & + & a_{22}x_2 & + & a_{23}x_3 & + & \cdots & + & a_{2n}x_n & = & c_2 \\ \cdot & & \cdot & & \cdot & & & & \cdot & & \cdot \\ \cdot & & \cdot & & \cdot & & & & \cdot & & \cdot \\ \cdot & & \cdot & & \cdot & & & & \cdot & & \cdot \\ a_{n1}x_1 & + & a_{n2}x_2 & + & a_{n3}x_3 & + & \cdots & + & a_{nn}x_n & = & c_n \end{array} \tag{9-2}$$

We can show that the set of Equations (9-2) can be represented by the product of two matrices:

$$\begin{pmatrix} a_{11} & a_{12} & a_{13} & \cdots & a_{1n} \\ a_{21} & a_{22} & a_{23} & \cdots & a_{2n} \\ \cdot & \cdot & \cdot & & \cdot \\ \cdot & \cdot & \cdot & & \cdot \\ \cdot & \cdot & \cdot & & \cdot \\ a_{n1} & a_{n2} & a_{n3} & \cdots & a_{nn} \end{pmatrix} \begin{pmatrix} x_1 \\ x_2 \\ \cdot \\ \cdot \\ \cdot \\ x_n \end{pmatrix} = \begin{pmatrix} c_1 \\ c_2 \\ \cdot \\ \cdot \\ \cdot \\ c_n \end{pmatrix} \tag{9-3}$$

The determinant of the coefficients is

$$D = \begin{vmatrix} a_{11} & a_{12} & a_{13} & \cdots & a_{1n} \\ a_{21} & a_{22} & a_{23} & \cdots & a_{2n} \\ \cdot & \cdot & \cdot & & \cdot \\ \cdot & \cdot & \cdot & & \cdot \\ \cdot & \cdot & \cdot & & \cdot \\ a_{n1} & a_{n2} & a_{n3} & \cdots & a_{nn} \end{vmatrix}$$

From Rule 4 in the previous section, we can write

$$Dx_1 = \begin{vmatrix} a_{11}x_1 & a_{12} & a_{13} & \cdots & a_{1n} \\ a_{21}x_1 & a_{22} & a_{23} & \cdots & a_{2n} \\ \cdot & \cdot & \cdot & & \cdot \\ \cdot & \cdot & \cdot & & \cdot \\ \cdot & \cdot & \cdot & & \cdot \\ a_{n1}x_1 & a_{n2} & a_{n3} & \cdots & a_{nn} \end{vmatrix}$$

If we now multiply each element in column 2 by x_2, each element in column 3 by x_3, and so on, and add these to column 1 in determinant Dx_1, then, by Rule 5 in the previous section, Dx_1 remains unchanged. That is,

$$Dx_1 = \begin{vmatrix} a_{11}x_1 + a_{12}x_2 + \cdots + a_{1n}x_n & a_{12} & a_{13} & \cdots & a_{1n} \\ a_{21}x_1 + a_{22}x_2 + \cdots + a_{2n}x_n & a_{22} & a_{23} & \cdots & a_{2n} \\ \cdot & \cdot & \cdot & & \cdot \\ \cdot & \cdot & \cdot & & \cdot \\ \cdot & \cdot & \cdot & & \cdot \\ a_{n1}x_1 + a_{n2}x_2 + \cdots + a_{nn}x_n & a_{n2} & a_{n3} & \cdots & a_{nn} \end{vmatrix}$$

Substituting Equations (9-2) for column 1, we have

$$Dx_1 = \begin{vmatrix} c_1 & a_{12} & a_{13} & \cdots & a_{1n} \\ c_2 & a_{22} & a_{23} & \cdots & a_{2n} \\ \cdot & \cdot & \cdot & & \cdot \\ \cdot & \cdot & \cdot & & \cdot \\ \cdot & \cdot & \cdot & & \cdot \\ c_n & a_{n2} & a_{n3} & \cdots & a_{nn} \end{vmatrix} = D_1$$

By the same argument used for column 1, we can write $Dx_2 = D_2$, $Dx_3 = D_3$, and so on, where D_i is the determinant D in which the elements in column i have been replaced by $c_1\ c_2\ c_3 \ldots$, etc. Hence, if $D \neq 0$, we have the following solutions of the given set of equations:

$$x_1 = \frac{D_1}{D}, \quad x_2 = \frac{D_2}{D}, \quad x_3 = \frac{D_3}{D}, \quad \text{and so on.}$$

This method of solving sets of linear equations is known as *Cramer's rule.*

Example

Solve the following equations, using Cramer's rule:

$$\begin{aligned} x + y + z &= 2 \\ 3x + y - 2z &= -5 \\ 2x - y - 3z &= -5 \end{aligned}$$

Solution. We first set up the determinants:

$$D = \begin{vmatrix} 1 & 1 & 1 \\ 3 & 1 & -2 \\ 2 & -1 & -3 \end{vmatrix} = -5 \qquad D_1 = \begin{vmatrix} 2 & 1 & 1 \\ -5 & 1 & -2 \\ -5 & -1 & -3 \end{vmatrix} = -5$$

$$D_2 = \begin{vmatrix} 1 & 2 & 1 \\ 3 & -5 & -2 \\ 2 & -5 & -3 \end{vmatrix} = 10 \qquad D_3 = \begin{vmatrix} 1 & 1 & 2 \\ 3 & 1 & -5 \\ 2 & -1 & -5 \end{vmatrix} = -15$$

$$x = \frac{-5}{-5} = 1 \qquad y = \frac{10}{-5} = -2 \qquad z = \frac{-15}{-5} = 3$$

Consider, now, a special set of linear equations that are all equal to zero:

$$\begin{aligned} a_{11}x_1 + a_{12}x_2 + a_{13}x_3 + \cdots + a_{1n}x_n &= 0 \\ a_{21}x_1 + a_{22}x_2 + a_{23}x_3 + \cdots + a_{2n}x_n &= 0 \\ \cdot \quad \cdot \quad \cdot \quad \cdot \quad & \cdot \\ \cdot \quad \cdot \quad \cdot \quad \cdot \quad & \cdot \\ \cdot \quad \cdot \quad \cdot \quad \cdot \quad & \cdot \\ a_{n1}x_1 + a_{n2}x_2 + a_{n3}x_3 + \cdots + a_{nn}x_n &= 0 \end{aligned} \tag{9-4}$$

Writing these equations in matrix form, we have

$$\begin{pmatrix} a_{11} & a_{12} & a_{13} & \cdots & a_{1n} \\ a_{21} & a_{22} & a_{23} & \cdots & a_{2n} \\ \cdot & \cdot & \cdot & & \cdot \\ \cdot & \cdot & \cdot & & \cdot \\ \cdot & \cdot & \cdot & & \cdot \\ a_{n1} & a_{n2} & a_{n3} & & a_{nn} \end{pmatrix} \begin{pmatrix} x_1 \\ x_2 \\ \cdot \\ \cdot \\ \cdot \\ x_n \end{pmatrix} = 0$$

or

$$\mathbf{Ax} = 0$$

If we tried to solve this set of equations by using Cramer's rule, we would obtain a trivial set of solutions $x_1 = x_2 = x_3 = \ldots = x_n = 0$. If we wish a nontrivial set of solutions (i.e., a set of solutions for which $\mathbf{x}$ is not equal to zero), then $\mathbf{A}$ must equal zero That is, $\mathbf{A}$ must be a singular matrix. Let us explore this idea further by looking at the characteristic equation of a matrix.

9-5 CHARACTERISTIC EQUATION OF A MATRIX

Consider a set of linear equations, such as those described in Section 9-4, written in matrix form as

$$\mathbf{Ac} = \lambda \mathbf{c} \tag{9-5}$$

where λ is a set of scalar constants called the *eigenvalues* of matrix $\mathbf{A}$, and $\mathbf{c}$ is a column matrix—a vector (see Chapter 8) called the *eigenvector*—belonging to $\mathbf{A}$. The only effect that matrix $\mathbf{A}$ has on matrix $\mathbf{c}$ is to multiply each element of $\mathbf{c}$ by a scalar constant λ. It is easy to show, then, that $\mathbf{c}$ and $\lambda\mathbf{c}$ are parallel vectors in space. Thus, the constant factor λ changes the length of $\mathbf{c}$, but not its direction.

Equation (9-5) can be written as

$$(\mathbf{A} - \lambda)\mathbf{c} = 0 \tag{9-6}$$

We saw in Section 9-4, however, that in order not to obtain a trivial set of solutions of the linear equations represented by Equation (9-6) (i.e., a solution different from $c_1 = c_2 = c_3 = \ldots = 0$), the matrix $(\mathbf{A} - \lambda)$ must be singular [i.e., det $(\mathbf{A} - \lambda) = 0$]. Writing the determinant $(\mathbf{A} - \lambda)$ in explicit form, we have

$$\begin{vmatrix} a_{11} - \lambda_1 & a_{12} & \cdots & a_{1n} \\ a_{21} & a_{22} - \lambda_2 & \cdots & a_{2n} \\ \vdots & \vdots & & \vdots \\ a_{n1} & a_{n2} & \cdots & a_{nn} - \lambda_n \end{vmatrix} = 0 \tag{9-7}$$

The determinant expressed in Equation (9-7) is called the *secular determinant*, and the linear equations represented by Equation (9-6) are called the *secular equations*. We can solve this $n \times n$ determinant using the method of cofactors, which will yield an nth-order polynomial

$$\lambda_i^n + a_1\lambda_i^{n-1} + a_2\lambda_i^{n-2} + \cdots + a_{n-1}\lambda_i + a_n = 0 \tag{9-8}$$

Equation (9-8) is called the *characteristic equation* of the matrix $\mathbf{A}$. The roots of Equation (9-8) are called the *eigenvalue spectrum* of $\mathbf{A}$. For every eigenvalue λ_i, there exists a corresponding eigenvector $\mathbf{c}_i$. Therefore, if there are n eigenvalues, then there are n eigenvalue equations,

$$\begin{aligned} \mathbf{A}\mathbf{c}_1 &= \lambda_1\mathbf{c}_1 \\ \mathbf{A}\mathbf{c}_2 &= \lambda_2\mathbf{c}_2 \\ \mathbf{A}\mathbf{c}_3 &= \lambda_3\mathbf{c}_3 \\ &\vdots \\ \mathbf{A}\mathbf{c}_n &= \lambda_n\mathbf{c}_n \end{aligned} \tag{9-9}$$

where each vector $\mathbf{c}_i$ is a column matrix. Equations (9-9) can be represented by the single matrix equation

$$\mathbf{AC} = \mathbf{C\Lambda} \tag{9-10}$$

or

$$\begin{pmatrix} a_{11} & a_{12} & \cdots & a_{1n} \\ a_{21} & a_{22} & \cdots & a_{2n} \\ \vdots & \vdots & & \vdots \\ a_{n1} & a_{n2} & \cdots & a_{nn} \end{pmatrix} \begin{pmatrix} c_{11} & c_{12} & \cdots & c_{1n} \\ c_{21} & c_{22} & \cdots & c_{2n} \\ \vdots & \vdots & & \vdots \\ c_{n1} & c_{n2} & \cdots & c_{nn} \end{pmatrix}$$

$$= \begin{pmatrix} c_{11} & c_{12} & \cdots & c_{1n} \\ c_{21} & c_{22} & \cdots & c_{2n} \\ \vdots & \vdots & & \vdots \\ c_{n1} & c_{n2} & \cdots & c_{nn} \end{pmatrix} \begin{pmatrix} \lambda_1 & 0 & \cdots & 0 \\ 0 & \lambda_2 & \cdots & 0 \\ \vdots & \vdots & & \vdots \\ 0 & 0 & \cdots & \lambda_n \end{pmatrix}$$

If det $\mathbf{C} \neq 0$ (i.e., if $\mathbf{C}$ is nonsingular), then there exists an inverse $\mathbf{C}^{-1}$, and multiplying through Equation (9-10) by $\mathbf{C}^{-1}$ gives

$$\mathbf{C}^{-1}\mathbf{AC} = \mathbf{C}^{-1}\mathbf{C}\Lambda = \Lambda \tag{9-11}$$

We see, then, that we can diagonalize the matrix $\mathbf{A}$ by compounding the eigenvectors into a matrix. Finding the eigenvectors that diagonalize a matrix to give its eigenvalues is one of the major types of problems that appear in quantum mechanics. (The topic of eigenvalues and their relationship to operators is discussed further in Chapter 10.)

Example

A problem in simple Hückel molecular orbital theory requires putting the matrix describing the set of secular equations, an example of which is

$$\begin{pmatrix} \alpha - E & \beta & 0 \\ \beta & \alpha - E & \beta \\ 0 & \beta & \alpha - E \end{pmatrix} \begin{pmatrix} c_1 \\ c_2 \\ c_3 \end{pmatrix} = 0$$

in diagonal form in order to determine the eigenvalues E. (Note that while we write this matrix equation in the simple form shown, we understand that there will be three eigenvalues (roots) of the matrix, with a set of coefficients c_1, c_2, and c_3 for each eigenvalue.) In this problem, α and β are integrals, called the *Coulomb integral* and *resonance integral*, respectively. The problem differs somewhat from the general description given in that, in this case, at least some values of E may not be distinct. When eigenvalues are not all distinct, it may not be possible to put the matrix in true diagonal form. Moreover, in a Hückel molecular orbital problem, not all of the secular equations are independent. This presents a special problem with which we must deal; in the example shown here, however, all the eigenvalues *are* distinct. In order not to obtain a trivial solution to the secular equations (i.e., $c_1 = c_2 = c_3 = 0$), the square matrix must be singular. That is,

$$\begin{vmatrix} \alpha - E & \beta & 0 \\ \beta & \alpha - E & \beta \\ 0 & \beta & \alpha - E \end{vmatrix} = 0$$

This determinant is most easily solved if we divide through it by β^3 and let $(\alpha - E)/\beta = x$:

$$\beta^3 \begin{vmatrix} \dfrac{\alpha - E}{\beta} & 1 & 0 \\ 1 & \dfrac{\alpha - E}{\beta} & 1 \\ 0 & 1 & \dfrac{\alpha - E}{\beta} \end{vmatrix} = 0 = \begin{vmatrix} x & 1 & 0 \\ 1 & x & 1 \\ 0 & 1 & x \end{vmatrix}$$

Using the method of cofactors, we find that the characteristic equation of the matrix is

$$x^3 - 2x = 0$$

which has the roots $x = 0, \pm\sqrt{2}$. Each root corresponds to a specific eigenvalue (energy state in this case) with its own set of coefficients c_i: $E_1 = \alpha + \sqrt{2}\beta$, $E_2 = 0$, $E_3 = \alpha - \sqrt{2}\beta$. To find the coefficients associated with each energy state, we substitute each energy value, in turn, into the secular equations and solve for the coefficients. For example, substituting $E_1 = \alpha + \sqrt{2}\beta$ into the secular equations

$$\begin{array}{ccccccc} (\alpha - E)c_1 & + & \beta c_2 & & & = & 0 \\ \beta c_1 & + & (\alpha - E)c_2 & + & \beta c_3 & = & 0 \\ 0 & + & \beta c_2 & + & (\alpha - E)c_3 & = & 0 \end{array}$$

gives $c_1 = (1/\sqrt{2})c_2$ and $c_1 = c_3$. Note that because the three secular equations are not independent, we can determine only the *ratio* of the coefficients and not their values. We essentially have three unknowns, but only two equations. Thus, we need another equation. If the eigenvectors are normalized, then we have another equation connecting the coefficients, namely,

$$\sum c_i^2 = 1$$

Therefore, in this example, $c_1^2 + c_2^2 + c_3^2 = 1$. Substituting the given values into this equation gives $c_1 = c_3 = 1/2$ and $c_2 = \sqrt{2}/2$. Coefficients for the other eigenvalues can be found in a similar manner.

SUGGESTED READINGS

1. NAGLE, R. KENT, and SAFF, EDWARD B. *Fundamentals of Differential Equations and Boundary Value Problems*, 2nd ed. Boston: Addison-Wesley, 1996.
2. SULLIVAN, MICHAEL. *College Algebra*, 4th ed. Upper Saddle River, NJ: Prentice Hall, 1996.
3. WASHINGTON, ALLYN J. *Basic Technical Mathematics*, 6th ed. Boston: Addison-Wesley, 1995.

PROBLEMS

1. Evaluate the following determinants:

(a) $\begin{vmatrix} 1 & 2 \\ 3 & 4 \end{vmatrix}$ **(b)** $\begin{vmatrix} 6 & 1 \\ -1 & -1 \end{vmatrix}$ **(c)** $\begin{vmatrix} 4 & -3 \\ 0 & 1 \end{vmatrix}$

(d) $\begin{vmatrix} 1 & 0 \\ 0 & 1 \end{vmatrix}$ **(e)** $\begin{vmatrix} x & 1 \\ 1 & x \end{vmatrix}$ **(f)** $\begin{vmatrix} \sec\theta & \tan\theta \\ \tan\theta & \sec\theta \end{vmatrix}$

(g) $\begin{vmatrix} 1 & 2 & 3 \\ 3 & 0 & 1 \\ -1 & 4 & 2 \end{vmatrix}$ **(h)** $\begin{vmatrix} 4 & 2 & 1 \\ -1 & 6 & 3 \\ -1 & 5 & -1 \end{vmatrix}$ **(i)** $\begin{vmatrix} x & 1 & 0 \\ 1 & x & 1 \\ 0 & 1 & x \end{vmatrix}$

(j) $\begin{vmatrix} 4 & 3 & 1 & -1 \\ 6 & 1 & 0 & -3 \\ 1 & 5 & 2 & -2 \\ 8 & 6 & -5 & 0 \end{vmatrix}$ **(k)** $\begin{vmatrix} x & b & 0 & 0 \\ b & x & b & 0 \\ 0 & b & x & b \\ 0 & 0 & b & x \end{vmatrix}$

2. Solve the following determinants for x:

(a) $\begin{vmatrix} x & 1 \\ 1 & x \end{vmatrix} = 0$ **(b)** $\begin{vmatrix} x & -2 \\ 1 & x \end{vmatrix} = 6$ **(c)** $\begin{vmatrix} 2x & 4 \\ 2 & x \end{vmatrix} = 2$

(d) $\begin{vmatrix} x & 1 & 1 \\ 1 & x & 1 \\ 1 & 1 & x \end{vmatrix} = 2$ **(e)** $\begin{vmatrix} x & 1 & 1 & 1 \\ 1 & x & 0 & 0 \\ 1 & 0 & x & 0 \\ 1 & 0 & 0 & x \end{vmatrix} = 0$

3. Add the matrices:

$$\begin{pmatrix} 1 & 1 & 4 & 3 \\ -1 & 0 & 1 & 2 \\ -1 & 2 & 4 & -3 \\ 5 & 6 & 3 & 5 \end{pmatrix} + \begin{pmatrix} 4 & 0 & -4 & 3 \\ 6 & 3 & -7 & 5 \\ -1 & 1 & -1 & 0 \\ -5 & 2 & 6 & 7 \end{pmatrix}$$

4. Perform the following matrix multiplication:

(a) $\begin{pmatrix} 1 & 4 \\ 3 & 2 \end{pmatrix}\begin{pmatrix} 6 & -3 \\ -3 & 1 \end{pmatrix}$ **(b)** $\begin{pmatrix} 1 & 0 \\ 0 & 1 \end{pmatrix}\begin{pmatrix} 4 & -1 \\ 2 & 3 \end{pmatrix}$

(c) $\begin{pmatrix} 3 & 0 & 3 \\ 4 & -1 & -1 \\ 1 & 2 & 5 \end{pmatrix}\begin{pmatrix} 1 & 1 & 1 \\ -2 & 1 & 6 \\ 3 & 4 & 5 \end{pmatrix}$

(d) $\begin{pmatrix} 1 & 4 & 1 \\ 0 & 1 & 2 \\ 2 & 4 & -3 \end{pmatrix}\begin{pmatrix} 0 & -4 & 3 \\ 6 & 3 & 7 \\ 2 & 6 & 7 \end{pmatrix}$ **(e)** $\begin{pmatrix} 1 & 8 & 4 \\ -2 & 3 & 0 \\ 5 & -1 & -1 \end{pmatrix}\begin{pmatrix} x \\ y \\ z \end{pmatrix}$

5. Given the matrices

$$\mathbf{A} = \begin{pmatrix} 1 & 1 & 4 \\ 2 & -6 & 10 \\ 4 & -1 & -1 \end{pmatrix} \quad \text{and} \quad \mathbf{B} = \begin{pmatrix} 6 & 1 & 0 \\ 4 & 2 & -1 \\ 8 & -4 & 3 \end{pmatrix}$$

show that $\mathbf{AB} \neq \mathbf{BA}$.

6. Solve the following sets of equations, using Cramer's rule:

(a)
$$\begin{aligned} x + y &= 3 \\ 4x - 3y &= 5 \end{aligned}$$

(b)
$$\begin{aligned} x + 2y + 3z &= -5 \\ -x - 3y + z &= -14 \\ 2x + y + z &= 1 \end{aligned}$$

(c)
$$\begin{aligned} x + 2y - z + t &= 2 \\ x - 2y + z - 3t &= 6 \\ 2x + y + 2z + t &= -4 \\ 3x + 3y + z - 2t &= 10 \end{aligned}$$

(d)
$$\begin{aligned} x \sin\theta + y\cos\theta &= x' \\ -x\cos\theta + y\sin\theta &= y' \end{aligned}$$

7. Show that only a trivial solution is possible for the following set of equations:

$$\begin{aligned} x + y &= 0 \\ x - y &= 0 \end{aligned}$$

8. Show that the matrix

$$\mathbf{E} = \begin{pmatrix} 1 & 0 & 0 \\ 0 & 1 & 0 \\ 0 & 0 & 1 \end{pmatrix}$$

will transform the vector $\begin{pmatrix} x \\ y \\ z \end{pmatrix}$ into itself.

9. Show that the matrix

$$\mathbf{C}_2 = \begin{pmatrix} -1 & 0 & 0 \\ 0 & -1 & 0 \\ 0 & 0 & 1 \end{pmatrix}$$

will transform the vector $\begin{pmatrix} x \\ y \\ z \end{pmatrix}$ into $\begin{pmatrix} -x \\ -y \\ z \end{pmatrix}$.

10. Prove that the inverse of the matrix

$$\mathbf{C}_3 = \begin{pmatrix} -\frac{1}{2} & \frac{\sqrt{3}}{2} & 0 \\ -\frac{\sqrt{3}}{2} & -\frac{1}{2} & 0 \\ 0 & 0 & 1 \end{pmatrix}$$

is

$$\mathbf{C}_3^{-1} = \begin{pmatrix} -\frac{1}{2} & -\frac{\sqrt{3}}{2} & 0 \\ \frac{\sqrt{3}}{2} & -\frac{1}{2} & 0 \\ 0 & 0 & 1 \end{pmatrix}$$

11. Find the inverse of the following matrix:

$$\begin{pmatrix} 1 & 0 & -1 \\ 3 & 2 & 4 \\ -2 & 1 & 0 \end{pmatrix}$$

12. Put the following matrix in diagonal form:

$$\mathbf{A} = \begin{pmatrix} 1 & \sqrt{6} \\ \sqrt{6} & 2 \end{pmatrix}$$

13. Show that the eigenvectors

$$\mathbf{C} = \begin{pmatrix} \frac{\sqrt{6}}{\sqrt{15}} & \frac{\sqrt{6}}{\sqrt{10}} \\ \frac{3}{\sqrt{15}} & \frac{-2}{\sqrt{10}} \end{pmatrix}$$

will diagonalize matrix $\mathbf{A}$ in Problem 12. (*Hint*: Show by matrix multiplication that $\mathbf{C}^{-1}\mathbf{A}\mathbf{C} = \Lambda$, where Λ is the diagonal form of $\mathbf{A}$.)

14. Show that $\mathbf{c}$ and $\lambda\mathbf{c}$, where λ is a scalar, are parallel vectors in space.

15. Show that the matrix of coefficients

$$\mathbf{C} = \begin{pmatrix} \frac{1}{2} & \frac{\sqrt{2}}{2} & \frac{1}{2} \\ \frac{\sqrt{2}}{2} & 0 & -\frac{\sqrt{2}}{2} \\ \frac{1}{2} & -\frac{\sqrt{2}}{2} & \frac{1}{2} \end{pmatrix}, \quad \text{where } \mathbf{C}^{-1} = \begin{pmatrix} \frac{1}{2} & \frac{\sqrt{2}}{2} & \frac{1}{2} \\ \frac{\sqrt{2}}{2} & 0 & -\frac{\sqrt{2}}{2} \\ \frac{1}{2} & -\frac{\sqrt{2}}{2} & \frac{1}{2} \end{pmatrix}$$

will put the matrix

$$\begin{pmatrix} \alpha - E & \beta & 0 \\ \beta & \alpha - E & \beta \\ 0 & \beta & \alpha - E \end{pmatrix}$$

into diagonal form and that the diagonal elements are the roots of the characteristic equation.

16. Solve the following set of secular equations for E in terms of α and β:

$$\begin{aligned} (\alpha - E)c_1 + \beta c_2 + \beta c_3 &= 0 \\ \beta c_1 + (\alpha - E)c_2 + \beta c_3 &= 0 \\ \beta c_1 + \beta c_2 + (\alpha - E)c_3 &= 0 \end{aligned}$$

Determine the relationship between the coefficients c_1, c_2, and c_3 for each value of E, and, using the fact that $\sum c_i^2 = 1$ (i.e., the eigenvectors must be normalized) for each energy state, find the values of c_1, c_2, and c_3 for each energy state.

CHAPTER 10

Operators

10-1 INTRODUCTION

An *operator* is a symbol that designates a process that transforms one function into another function. For example, the operator $\hat{\mathbf{D}}_x = \partial/\partial x$, applied to the function $f(x)$, designates that the first derivative of $f(x)$ with respect to x is to be taken. Following are some other examples of operators:

$$\begin{aligned}\Delta:&\quad \Delta f(x) = f(x+h) - f(x)\\ \Sigma:&\quad \Sigma f(x) = f_1(x) + f_2(x) + f_3(x) + \cdots\\ \Pi:&\quad \Pi f(x) = f_1(x)\cdot f_2(x)\cdot f_3(x)\cdot \ldots\end{aligned}$$

If two or more operators are applied simultaneously to a function, then the operator immediately adjacent to the function operates on the function first, giving a new function; the next adjacent operator operates on this new function, giving another new function, and so on. For example, consider, on the one hand, the application of the operators $\hat{\mathbf{D}}_x = \partial/\partial x$ and $\hat{\mathbf{D}}_y = \partial/\partial y$ to the function $f(x, y) = x^3y^2$:

$$\hat{\mathbf{D}}_x\hat{\mathbf{D}}_y(x^3y^2) = \hat{\mathbf{D}}_x(2x^3y) = 6x^2y$$

On the other hand,

$$\hat{\mathbf{D}}_y\hat{\mathbf{D}}_x(x^3y^2) = \hat{\mathbf{D}}_y(3x^2y^2) = 6x^2y$$

When the result of the operations is independent of the order in which the operators are applied, as in the foregoing example, the operators are said to *commute.* That is,

$$\hat{\mathbf{D}}_x\hat{\mathbf{D}}_y f(x, y) = \hat{\mathbf{D}}_y\hat{\mathbf{D}}_x f(x, y)$$

or

$$[\hat{\mathbf{D}}_x\hat{\mathbf{D}}_y - \hat{\mathbf{D}}_y\hat{\mathbf{D}}_x]f(x, y) = 0 \tag{10-1}$$

Here, the brackets [ ] are known as *commutator brackets*, and the expression in the brackets, itself an operator, is called the *commutator of the operators*. It is important to note that the function $f(x, y)$ has not been algebraically factored out of the left-hand side of Equation (10-1). Equation (10-1) is an operator equation and signifies that when the operator $[\hat{\mathbf{D}}_x\hat{\mathbf{D}}_y - \hat{\mathbf{D}}_y\hat{\mathbf{D}}_x]$ acts on the function $f(x, y)$, it produces the number zero.

In general, then, we can say that two operators $\hat{\mathbf{A}}$ and $\hat{\mathbf{B}}$ commute if, and only if, [1]

$$[\hat{\mathbf{A}}\hat{\mathbf{B}} - \hat{\mathbf{B}}\hat{\mathbf{A}}]f(x) = 0 \tag{10-2}$$

If the commutator does not equal zero, then the order in which the operators are applied will affect the results and must be taken into account.

Certain operators, such as differential operators, may be applied to a function more than once. For example, applying the operator $\hat{\mathbf{D}}_x = \partial/\partial x$ twice to the function $f(x)$ yields the second derivative of the function with respect to x:

$$\hat{\mathbf{D}}_x\hat{\mathbf{D}}_x f(x) = \hat{\mathbf{D}}_x^2 f(x) = \frac{\partial}{dx}\left(\frac{\partial f}{\partial x}\right) = \frac{\partial^2 f}{\partial x^2}$$

A certain amount of care is required in applying an operator to a function more than once, particularly if the operator contains more than one term. For example, consider the operator $\partial/\partial x$ in plane polar coordinates:

$$\hat{\mathbf{D}}_x = \frac{\partial}{\partial x} = \cos\theta\frac{\partial}{\partial r} - \frac{\sin\theta}{r}\frac{\partial}{\partial \theta}$$

The operation $\hat{\mathbf{D}}_x^2\phi$ simply means *apply the operator* $\hat{\mathbf{D}}_x$ *to the function twice*. The first application yields

$$\hat{\mathbf{D}}_x f(r, \theta) = \left(\cos\theta\frac{\partial f}{\partial r} - \frac{\sin\theta}{r}\frac{\partial f}{\partial \theta}\right)$$

The second application is a little tricky, since some terms must be differentiated as a product. We get

$$\hat{\mathbf{D}}_x^2 f(r, \theta) = \left(\cos\theta\frac{\partial}{\partial r} - \frac{\sin\theta}{r}\frac{\partial}{\partial \theta}\right)\left(\cos\theta\frac{\partial f}{\partial r} - \frac{\sin\theta}{r}\frac{\partial f}{\partial \theta}\right)$$

Each term in the $\hat{\mathbf{D}}_x$ operator operates on both terms in $\hat{\mathbf{D}}_x f(r, \theta)$:

$$\hat{\mathbf{D}}_x^2 f(r, \theta) = \cos\theta\frac{\partial}{\partial r}\left(\cos\theta\frac{\partial f}{\partial r}\right) + \cos\theta\frac{\partial}{\partial r}\left(-\frac{\sin\theta}{r}\frac{\partial f}{\partial \theta}\right)$$

[1] Equation (10-2) may be written in an abbreviated form, $[\hat{\mathbf{A}}\hat{\mathbf{B}} - \hat{\mathbf{B}}\hat{\mathbf{A}}] = 0$, or even $[\hat{\mathbf{A}}, \hat{\mathbf{B}}] = 0$, provided that it is understood what the abbreviation implies.

$$
-\frac{\sin\theta}{r}\frac{\partial}{\partial\theta}\left(\cos\theta\frac{\partial f}{\partial r}\right)-\frac{\sin\theta}{r}\frac{\partial}{\partial\theta}\left(-\frac{\sin\theta}{r}\frac{\partial f}{\partial\theta}\right)
$$

$$
=\cos^2\theta\frac{\partial}{\partial r}\left(\frac{\partial f}{\partial r}\right)-\sin\theta\cos\theta\frac{\partial}{\partial r}\left(\frac{1}{r}\frac{\partial f}{\partial\theta}\right)
$$

$$
-\frac{\sin\theta}{r}\frac{\partial}{\partial\theta}\left(\cos\theta\frac{\partial f}{\partial r}\right)+\frac{\sin\theta}{r^2}\frac{\partial}{\partial\theta}\left(\sin\theta\frac{\partial f}{\partial\theta}\right)
$$

$$
=\cos^2\theta\frac{\partial^2 f}{\partial r^2}-\sin\theta\cos\theta\left(\frac{1}{r}\frac{\partial^2 f}{\partial r\,\partial\theta}-\frac{1}{r^2}\frac{\partial f}{\partial\theta}\right)
$$

$$
-\frac{\sin\theta}{r}\left(\cos\theta\frac{\partial^2 f}{\partial\theta\,\partial r}-\sin\theta\frac{\partial f}{\partial r}\right)+\frac{\sin\theta}{r^2}\left(\sin\theta\frac{\partial^2 f}{\partial\theta^2}+\cos\theta\frac{\partial f}{\partial\theta}\right)
$$

Note that while, in the first term of the preceding equation, $\cos\theta$ can pass through the $\partial/\partial r$ operator, giving $\cos^2\theta(\partial^2 f/\partial r^2)$, other terms, such as $\frac{\partial}{\partial r}(\frac{1}{r}\frac{\partial f}{\partial\theta})$, must be differentiated as a product, since both $1/r$ and $f(r,\theta)$ are functions of r. Therefore, collecting terms, while recognizing the commutative properties of mixed second derivatives, we have

$$
\hat{\mathbf{D}}_x^2 f(r,\theta)=\cos^2\theta\frac{\partial^2 f}{\partial\theta^2}-\frac{2\sin\theta\cos\theta}{r}\frac{\partial^2 f}{\partial r\,\partial\theta}+\frac{2\sin\theta\cos\theta}{r^2}\frac{\partial f}{\partial\theta}
$$

$$
+\frac{\sin^2\theta}{r}\frac{\partial f}{\partial r}+\frac{\sin^2\theta}{r^2}\frac{\partial^2 f}{\partial\theta^2}
$$

10-2 VECTOR OPERATORS

Recall from basic courses in physics that the components of force in the x, y, and z directions, namely, F_x, F_y, and F_z, are related to the potential energy $V(x, y, z)$ by the equations

$$
F_x=-\frac{\partial}{\partial x}V(x,y,z)
$$

$$
F_y=-\frac{\partial}{\partial y}V(x,y,z)
$$

$$
F_z=-\frac{\partial}{\partial z}V(x,y,z)
$$

Since the total force is a vector, that is,

$$\mathbf{F} = \mathbf{i}F_x + \mathbf{j}F_y + \mathbf{k}F_z$$

we have

$$\mathbf{F} = -\left(\mathbf{i}\frac{\partial V}{\partial x} + \mathbf{j}\frac{\partial V}{\partial y} + \mathbf{k}\frac{\partial V}{\partial z}\right) \tag{10-3}$$

Equation (10-3), however, can be written in operator form as

$$\mathbf{F} = -\left(\mathbf{i}\frac{\partial}{\partial x} + \mathbf{j}\frac{\partial}{\partial y} + \mathbf{k}\frac{\partial}{\partial z}\right)V = -\nabla V$$

where ∇V (read "del V") is known as the *gradient* of the scalar V and

$$\nabla = \mathbf{i}\frac{\partial}{\partial x} + \mathbf{j}\frac{\partial}{\partial y} + \mathbf{k}\frac{\partial}{\partial z} \tag{10-4}$$

is known as the *gradient operator*. Note that V in this example is a scalar, but ∇ operating on V produces the vector ∇V. Hence, force is the negative gradient of the potential energy.

It is possible to use ∇ in a scalar product with another vector $\mathbf{A}$. Such a product $\nabla \cdot \mathbf{A}$ is a scalar called the *divergence* of $\mathbf{A}$. If ϕ is any scalar, we find that the divergence of the gradient of ϕ, in Cartesian coordinates, is

$$\nabla \cdot \nabla\phi = \nabla^2\phi = \frac{\partial^2\phi}{\partial x^2} + \frac{\partial^2\phi}{\partial y^2} + \frac{\partial^2\phi}{\partial z^2} \tag{10-5}$$

The scalar operator

$$\nabla^2 = \frac{\partial^2}{\partial x^2} + \frac{\partial^2}{\partial y^2} + \frac{\partial^2}{\partial z^2} \tag{10-6}$$

(read "del squared") is known as the *Laplacian operator.*

Example

The total energy of a system, in terms of its momentum, is

$$E = \frac{p^2}{2m} + V = \frac{1}{2m}(p_x^2 + p_y^2 + p_z^2) + V$$

where p_q is the momentum, m is the mass, and V is the potential energy of the system. Making the appropriate quantum mechanical operator substitutions $p_q \rightarrow (h/2\pi i\, \partial/\partial q)$ and $V \rightarrow \hat{\mathbf{V}}$, derive the energy operator $\hat{\mathbf{H}}(i = \sqrt{-1})$.

Solution. The square of the momentum operator $p_q \rightarrow (h/2\pi i\, \partial/\partial q)$ is a second-derivative operator $p_q^2 \rightarrow -(h^2/4\pi^2)(\partial^2/\partial q^2)$. Therefore,

$$\hat{\mathbf{H}} = \frac{-h^2}{8\pi^2 m}\left(\frac{\partial^2}{\partial x^2} + \frac{\partial^2}{\partial y^2} + \frac{\partial^2}{\partial z^2}\right) + \hat{\mathbf{V}} = -\frac{h^2}{8\pi^2 m}\nabla^2 + \hat{\mathbf{V}}$$

$\hat{\mathbf{H}}$ is known as the *Hamiltonian operator.*

10-3 EIGENVALUE EQUATIONS REVISITED

In Chapter 9, we considered a matrix description of a set of linear equations written

$$\mathbf{Ac} = \lambda \mathbf{c}$$

where λ is a set of constants called the eigenvalues of the matrix $\mathbf{A}$ and $\mathbf{c}$ is a vector. We recognize the matrix $\mathbf{A}$ to be an operator, since its effect is to multiply each element in the matrix $\mathbf{c}$ by the constant λ. We observe that eigenvalue equations are not restricted to matrix mathematics, but find general use in many areas of applied mathematics. In fact, we can define as an *eigenvalue equation* any operator equation having the general form

$$\hat{\mathbf{A}}\phi = a\phi \tag{10-7}$$

where $\hat{\mathbf{A}}$ is an operator whose operation on a function ϕ, called an *eigenfunction* of the operator, produces a set of constants, the *eigenvalues*, multiplied by ϕ. For example, the differential equation

$$\frac{d}{dx}(e^{mx}) = me^{mx}$$

is an eigenvalue equation.

It is possible for operators to have several eigenfunctions; for example, the functions e^{mx}, e^{-mx}, e^{imx}, e^{-imx}, $\sin mx$, and $\cos mx$ are all eigenfunctions of the operator $\hat{\mathbf{D}}_x^2 = \partial^2/\partial x^2$. Some such functions, however, may be unsuitable in a problem because of certain restrictions known as *boundary conditions*.

We can solve eigenvalue equations by methods outlined for differential and partial differential equations in Chapter 7. Consider the following examples:

Examples

(a) Show that the functions $\phi = e^{-ar}$ are eigenfunctions of the operator $\hat{\mathbf{D}}^2 = \partial^2/\partial r^2$. What are the eigenvalues?

Solution. If $\phi = e^{-ar}$ are eigenfunctions of $\hat{\mathbf{D}}^2$, then they should satisfy the eigenvalue equation

$$\hat{\mathbf{D}}^2\phi = k\phi$$

where the constants k are the eigenvalues. Differentiating ϕ with respect to r, we have

$$\hat{\mathbf{D}}\phi = -ae^{-ar}$$

$$\hat{\mathbf{D}}^2\phi = a^2e^{-ar} = a^2\phi$$

Thus, we see that the functions $\phi = e^{-ar}$ satisfy the eigenvalue equation and that $k = a^2$ are the eigenvalues.

(b) The Schrödinger equation in one-dimension is

$$\frac{d^2\psi}{dx^2} + \frac{8\pi^2m}{h^2}(E - V)\psi = 0$$

where h, m, and E are constants. Show that the Schrödinger equation is an eigenvalue equation. What is the operator and what are the eigenvalues? (*Hint*: In the preceding equation, d^2/dx^2 and V are both operators.)

Solution. Let us begin by putting the Schrödinger equation in eigenvalue form:

$$\frac{d^2\psi}{dx^2} + \frac{8\pi^2mE}{h^2}\psi - \frac{8\pi^2mV}{h^2}\psi = 0$$

$$-\frac{h^2}{8\pi^2m}\frac{d^2\psi}{dx^2} - E\psi + V\psi = 0$$

$$\left(-\frac{h^2}{8\pi^2m}\frac{d^2}{dx^2} + V\right)\psi = E\psi$$

We recognize the operator in parentheses to be $\hat{\mathbf{H}}$, the Hamiltonian operator in one dimension. (See the example in Section 10-2.) The eigenvalues of the Hamiltonian operator are the energy states E.

The Hamiltonian operator illustrates the following general property of operators: Let $\hat{\mathbf{D}}$ be the sum of two or more operators; that is, $\hat{\mathbf{D}} = \hat{\mathbf{D}}_1 + \hat{\mathbf{D}}_2 + \hat{\mathbf{D}}_3 + \cdots$. Then $\hat{\mathbf{D}}f(x) = (\hat{\mathbf{D}}_1 + \hat{\mathbf{D}}_2 + \hat{\mathbf{D}}_3 + \cdots)f(x) = \hat{\mathbf{D}}_1 f(x) + \hat{\mathbf{D}}_2 f(x) + \hat{\mathbf{D}}_3 f(x) + \cdots$.

10-4 HERMITIAN OPERATORS

Many eigenvalue problems found in physical chemistry—particularly those in the area of quantum mechanics—can be described by an eigenvalue equation having the general form

$$L(u) + \lambda w u = 0 \tag{10-8}$$

Generally, the operator L is a second-order differential operator that operates on the function u, the constants λ are the eigenvalues, and w is a weighting factor. Second-order differential equations of the form

$$\hat{\mathbf{L}}u = f\frac{d^2u}{dx^2} + g\frac{du}{dx} + hu$$

where f, g, and h are functions of x, are said to be *self-adjoint* if $g = df/dx$. If the differential operator is not self-adjoint, it can be made self-adjoint by multiplying it by a factor $\int((g - (df/dx))/f)\,dx$. For example, Legendre's equation is already self-adjoint. Hermite's equation can be made self-adjoint by multiplying it by e^{-x^2}. Self-adjoint operators, such as those just described, are found to obey the following important rule: If u and v are two acceptable functions of q, then an operator $\hat{\mathbf{P}}$ is said to be *Hermitian* if

$$\int_{\text{all space}} u^* \hat{\mathbf{P}} v \, d\tau = \int_{\text{all space}} v \hat{\mathbf{P}}^* u^* \, d\tau \tag{10-9}$$

where the * denotes the complex conjugate.

Example

Show that the momentum operator $\hat{\mathbf{p}}_q = (h/2\pi i)(\partial/\partial q)$ is Hermitian.

$$\int u^* \hat{\mathbf{p}}_q v \, dq = \frac{h}{2\pi i}\int u^* \frac{\partial v}{\partial q} dq$$

Solution. Integrating the right-hand side of the equation by parts, we have

$$\frac{h}{2\pi i}\int u^*\frac{\partial v}{\partial q}\,dq = \frac{h}{2\pi i}\left\{[u^*v]_{\text{endpoints}} - \int v\frac{\partial u^*}{\partial q}\,dq\right\}$$

We assume that u and v vanish at the endpoints. Thus, the expression in brackets is zero, which gives

$$\int u^*\hat{\mathbf{p}}_q v\,dq = -\frac{h}{2\pi i}\int v\frac{\partial u^*}{\partial q}\,dq = +\int v\hat{\mathbf{p}}_q{}^*u^*\,dq$$

The operator $\hat{\mathbf{p}}_q$ is Hermitian.

A very important property of Hermitian operators is that their eigenvalues are always real. To see this, consider the eigenvalue equation

$$\hat{\mathbf{p}}\psi = p\psi \tag{10-10}$$

where $\hat{\mathbf{p}}$ is a Hermitian operator. Therefore, it also must be true that

$$\hat{\mathbf{p}}^*\psi^* = p^*\psi^* \tag{10-11}$$

We now multiply Equation (10-10) by ψ^* and Equation (10-11) by ψ and integrate over all space:

$$\int \psi^*\hat{\mathbf{p}}\psi\,d\tau = p\int \psi^*\psi\,d\tau$$

$$\int \psi\,\hat{\mathbf{p}}^*\psi^*\,d\tau = p^*\int \psi^*\psi\,d\tau$$

Since

$$\int \psi^*\hat{\mathbf{p}}\psi\,d\tau = \int \psi\hat{\mathbf{p}}^*\psi^*\,d\tau$$

it must be true that $p = p^*$; hence, the eigenvalues are real.

10-5 ROTATIONAL OPERATORS

We saw in a previous section that operators can take matrix form. In the eigenvalue problem, the matrix operator affects the magnitude or length of a vector, but not its direction. Consider, now, another type of matrix operator that changes the direction of a vector, but not its magnitude or length. Such an operator is called a *rotational operator*, since its effect is to rotate a vector about the origin of a coordinate system through an angle θ. As we shall see, it does this by, in essence, rotating the coordinate system back through the angle θ.

Consider, as an example, a vector **r** extending from the origin of a Cartesian coordinate system to the point (x_1, y_1), as shown in Fig. 10-1(a). Suppose that we wish to rotate this vector through an angle θ to another point (x_2, y_2). We can do this either by rotating the vector itself through the angle θ [Fig. 10-1(b)] or by rotating the coordinate system back through the angle θ [Fig. 10-1(c)].

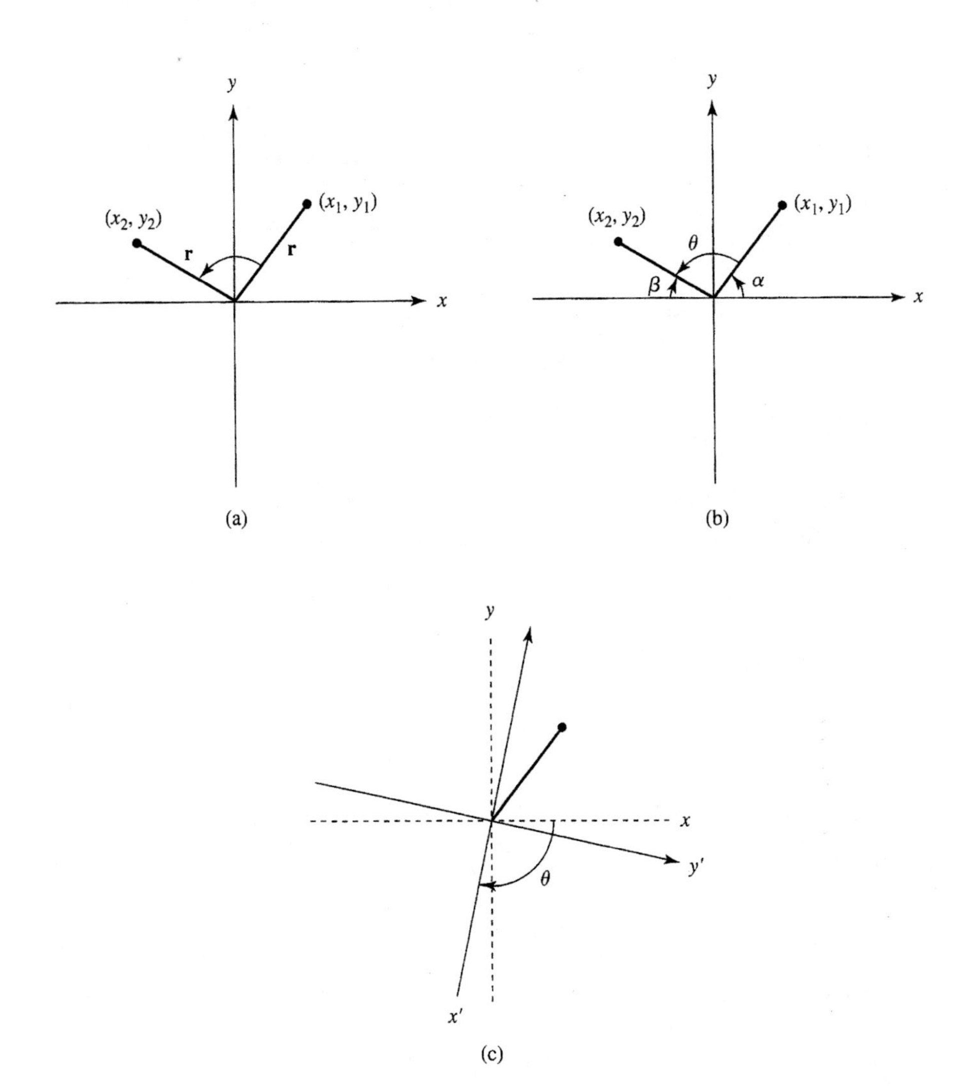

Figure 10-1 Rotation of a vector through an angle θ.

The coordinates (x_1, y_1) can be related to the coordinates (x_2, y_2) by simple trigonometry. We see from Fig. 10-1(b) that

$$\begin{aligned}
\theta &= 180^\circ - (\alpha + \beta) = 180^\circ - \alpha - \beta \\
x_1 &= r \cos \alpha \\
y_1 &= r \sin \alpha \\
x_2 &= -r \cos \beta \\
y_2 &= r \sin \beta \\
\beta &= 180^\circ - (\theta + \alpha)
\end{aligned}$$

We now must make use of some important trigonometric identities:

$$\begin{aligned}
\sin(A + B) &= \sin A \cos B + \cos A \sin B \\
\sin(A - B) &= \sin A \cos B - \cos A \sin B \\
\cos(A + B) &= \cos A \cos B - \sin A \sin B \\
\cos(A - B) &= \cos A \cos B + \sin A \sin B
\end{aligned} \tag{10-12}$$

Hence,

$$\begin{aligned}
\sin \beta &= \sin[180^\circ - (\theta + \alpha)] = \sin 180^\circ \cos(\theta + \alpha) - \cos 180^\circ \sin(\theta + \alpha) \\
&= \sin(\theta + \alpha) = \sin \theta \cos \alpha + \cos \theta \sin \alpha
\end{aligned}$$

However,

$$y_2 = r \sin \beta = \sin \theta (r \cos \alpha) + \cos \theta (r \sin \alpha)$$

Therefore,

$$y_2 = x_1 \sin \theta + y_1 \cos \theta$$

By the same method, we can show that

$$x_2 = x_1 \cos \theta - y_1 \sin \theta$$

These two equations can be written in matrix form as

$$\begin{pmatrix} x_2 \\ y_2 \end{pmatrix} = \begin{pmatrix} \cos \theta & -\sin \theta \\ \sin \theta & \cos \theta \end{pmatrix} \begin{pmatrix} x_1 \\ y_1 \end{pmatrix} \tag{10-13}$$

where we recognize $\begin{pmatrix} x_1 \\ y_1 \end{pmatrix}$ as the original vector and $\begin{pmatrix} x_2 \\ y_2 \end{pmatrix}$ as the rotated vector. The matrix

$$\begin{pmatrix} \cos\theta & -\sin\theta \\ \sin\theta & \cos\theta \end{pmatrix}$$

is a rotational operator, since it rotates a vector about the origin in two-dimensional space. This operator also can be thought of as a transformation operator, since it transforms the coordinates of a point (x_1, y_1) into a point (x_2, y_2) by a rotation of the coordinate axes back through an angle θ.

Examples

(a) Rotate the vector $\mathbf{r}\langle 3, 4\rangle$ through an angle of 180°, and find the new direction of the vector.

Solution. Since $\sin 180° = 0$ and $\cos 180° = -1$, we have, for the rotational operator,

$$\begin{pmatrix} \cos\theta & -\sin\theta \\ \sin\theta & \cos\theta \end{pmatrix} = \begin{pmatrix} -1 & 0 \\ 0 & -1 \end{pmatrix}$$

Allowing this operator to operate on the vector $\mathbf{r}\langle 3, 4\rangle$ gives

$$\begin{pmatrix} -1 & 0 \\ 0 & -1 \end{pmatrix}\begin{pmatrix} 3 \\ 4 \end{pmatrix} = \begin{pmatrix} -3 \\ -4 \end{pmatrix}$$

which produces a new vector, $\mathbf{r}\langle -3, -4\rangle$. It is easy to verify that the difference in direction between $\mathbf{r}\langle -3, -4\rangle$ and $\mathbf{r}\langle 3, 4\rangle$ is 180°.

(b) Find the new coordinates of the point (3, 4) after a rotation of the point about the origin by 90°.

Solution. Since $\sin 90° = 1$ and $\cos 90° = 0$, we have

$$\begin{pmatrix} \cos\theta & -\sin\theta \\ \sin\theta & \cos\theta \end{pmatrix} = \begin{pmatrix} 0 & -1 \\ 1 & 0 \end{pmatrix}$$

Allowing this operator to operate on the point (3, 4) gives

$$\begin{pmatrix} 0 & -1 \\ 1 & 0 \end{pmatrix}\begin{pmatrix} 3 \\ 4 \end{pmatrix} = \begin{pmatrix} -4 \\ 3 \end{pmatrix}$$

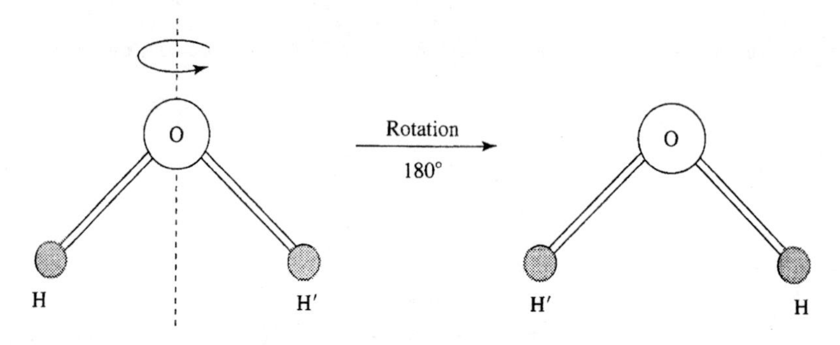

Figure 10-2 Rotation of a water molecule through an angle of 180°.

Rotational operators are useful in describing the symmetry of molecules. A designation known as Schoenflies notation uses the symbol $\mathbf{C}_n$ (where $n = 360/\theta$) to denote a particular rotational operator. For example, an operator that rotates a molecule through an angle of 90° is called a $\mathbf{C}_4$ operator. A molecule is said to possess $\mathbf{C}_4$ symmetry if a $\mathbf{C}_4$ operation leaves the molecule unchanged. Similar matrix operators for other types of molecular symmetry also are possible.

Consider the water molecule shown in Fig. 10-2. If we apply a $\mathbf{C}_2$ rotational operator to a water molecule so as to rotate the molecule through an angle of 180° about the axis shown in the figure (the dashed line), the molecule remains unchanged. (The hydrogen atoms are assumed to be indistinguishable.) We say, then, that the water molecule possesses a $\mathbf{C}_2$ axis of rotation. By assigning an x-, y-, and z-coordinate to each hydrogen atom (the oxygen is placed at the origin and the rotation axis is in the z direction), we can easily show that to be the case.

10-6 TRANSFORMATION OF ∇^2 TO PLANE POLAR COORDINATES

Many problems in classical and quantum mechanics require the use of the ∇^2 operator in coordinates other than Cartesian coordinates, the coordinate system in which the operator has its simplest form. In this section, we shall transform ∇^2 to plane polar coordinates. We choose plane polar coordinates because, in this transformation, all the essentials of transforming ∇^2 are demonstrated, while at the same time the transformation is not as lengthy as it would be to, say, spherical polar coordinates. (For those interested, the essentials of the transformation of ∇^2 to spherical polar coordinates are given in Appendix III.)

The Laplacian operator in two dimensions is

$$\nabla^2 = \frac{\partial^2}{\partial x^2} + \frac{\partial^2}{\partial y^2}$$

Recall from Chapter 1 that the transformation and reverse-transformation equations to plane polar coordinates are

$$x = r\cos\theta \qquad r = (x^2 + y^2)^{1/2}$$

$$y = r\sin\theta \qquad \tan\phi = \left(\frac{y}{x}\right)$$

The transformation of the first-derivative operators $\partial/\partial x$ and $\partial/\partial y$ can be found by the chain rule:

$$\frac{\partial}{\partial x} = \frac{\partial r}{\partial x}\frac{\partial}{\partial r} + \frac{\partial \theta}{\partial x}\frac{\partial}{\partial \theta} \quad \text{and} \quad \frac{\partial}{\partial y} = \frac{\partial r}{\partial y}\frac{\partial}{\partial r} + \frac{\partial \theta}{\partial y}\frac{\partial}{\partial \theta}$$

We now determine the transformation derivatives from the reverse-transformation equations:

$$\frac{\partial r}{\partial x} = \frac{1}{2}\left(x^2 + y^2\right)^{-1/2} 2x = \frac{x}{r} = \frac{r\cos\theta}{r} = \cos\theta$$

Likewise,

$$\frac{\partial r}{\partial y} = \sin\theta$$

Finding $\partial\theta/\partial x$ and $\partial\theta/\partial y$ is a little more involved. A little trick, to keep from having to differentiate the inverse tangent, is to differentiate $\tan\theta$ implicitly:

$$\frac{d}{dx}\tan\theta = \sec^2\theta\frac{d\theta}{dx} = -\frac{y}{x^2}$$

$$\frac{1}{\cos^2\theta}\frac{\partial\theta}{\partial x} = -\frac{r\sin\theta}{r^2\cos^2\theta}$$

$$\frac{\partial\theta}{\partial x} = -\frac{\sin\theta}{r}$$

Likewise, by the same procedure, we have

$$\frac{\partial\theta}{\partial y} = \frac{\cos\theta}{r}$$

The first derivatives are therefore

$$\frac{\partial}{\partial x} = \cos\theta\frac{\partial}{\partial r} - \frac{\sin\theta}{r}\frac{\partial}{\partial\theta} \quad \text{and} \quad \frac{\partial}{\partial y} = \sin\theta\frac{\partial}{\partial r} + \frac{\cos\theta}{r}\frac{\partial}{\partial\theta}$$

To find the second-derivative operators, we must now apply the operator twice on a function, taking care to watch for cases where terms must be differentiated as a product. The second-derivative operator $\partial^2/\partial x^2$ was derived in Section 10-1. Using the same method, we obtain, for $\partial^2/\partial y^2$,

$$\frac{\partial^2 f}{\partial y^2} = \sin^2\theta \frac{\partial^2 f}{\partial r^2} + \frac{2\sin\theta\cos\theta}{r}\frac{\partial^2 f}{\partial r\,\partial\theta} - \frac{2\sin\theta\cos\theta}{r^2}\frac{\partial f}{\partial\theta} + \frac{\cos^2\theta}{r}\frac{\partial f}{\partial r} + \frac{\cos^2\theta}{r^2}\frac{\partial^2 f}{\partial\theta^2}$$

Adding the two operators together gives

$$\nabla^2 = \frac{\partial^2}{\partial x^2} + \frac{\partial^2}{\partial y^2} = \frac{\partial^2}{\partial r^2} + \frac{1}{r}\frac{\partial}{\partial r} + \frac{1}{r^2}\frac{\partial^2}{\partial\theta^2}$$

SUGGESTED READING

1. NAGLE, R. KENT, and SAFF, EDWARD B. *Fundamentals of Differential Equations and Boundary Value Problems*, 2d ed. Boston: Addison-Wesley, 1996.

PROBLEMS

1. Perform the following operations:

   (a) $\sum_{n=0}^{5} x^n$

   (b) $\sum_{n=0}^{5} (-1)^n x^n$

   (c) ΔE

   (d) $\hat{\mathbf{D}}_x(x^3y)$, $\hat{\mathbf{D}}_x = \frac{\partial}{\partial x}$

   (e) $\hat{\mathbf{D}}_x^2(x^2y^3)$

   (f) $\hat{\mathbf{D}}_y\hat{\mathbf{D}}_x(x^4y^3)$

   (g) $\hat{\mathbf{D}}_z\hat{\mathbf{D}}_y\hat{\mathbf{D}}_x(x^2y^2z^2)$

   (h) $\hat{\mathbf{D}}_x \sum_{n=0}^{5} x^n$

   (i) $\prod_{n=0}^{4} x_n!$

   (j) $\begin{pmatrix} -1 & 0 \\ 0 & -1 \end{pmatrix}\begin{pmatrix} a \\ b \end{pmatrix}$

2. Determine whether the following pairs of operators commute:

   (a) $[\hat{\mathbf{D}}_y, \hat{\mathbf{D}}_z]$

   (b) $[\hat{\mathbf{D}}_x, \sum]$

   (c) $[\hat{\mathbf{D}}_x, \Delta]$

   (d) $[\sum, \sqrt{\ }]$

3. Show that $\nabla \cdot \nabla\phi = \frac{\partial^2\phi}{\partial x^2} + \frac{\partial^2\phi}{\partial y^2} + \frac{\partial^2\phi}{\partial z^2}$.

4. Show that $\nabla(\psi\phi) = \psi\,\nabla\phi + \phi\,\nabla\psi$.

5. An interpretation of the *Heisenberg uncertainty principle* is that the operator for linear momentum in the x direction does not commute with the operator for position along the x-axis. Let

$$\hat{\mathbf{p}}_x = \frac{h}{2\pi i}\frac{\partial}{\partial x} \quad \text{and} \quad \hat{\mathbf{x}} = x$$

where $i = \sqrt{-1}$ and h is a constant, represent operators for linear momentum and position along the x-axis, respectively. Evaluate the commutator

$$[\hat{\mathbf{p}}_x\hat{\mathbf{x}} - \hat{\mathbf{x}}\hat{\mathbf{p}}_x]$$

and show that it does not equal zero. (*Hint*: Apply the operators $\hat{\mathbf{x}}$ and $\hat{\mathbf{p}}_x$ to an arbitrary function $\phi(x)$, keeping in mind that $x\phi(x)$ must be differentiated as a product.)

6. Show that $y = \sin ax$ is not an eigenfunction of the operator d/dx, but is an eigenfunction of the operator d^2/dx^2.

7. Show that the function $\Phi = Ae^{im\phi}$, where i, m, and A are constants, is an eigenfunction of the angular momentum operator in the z-direction:

$$\hat{\mathbf{M}}_z = \frac{h}{2\pi i}\frac{\partial}{\partial \phi}$$

What are the eigenvalues?

8. Show that the function

$$\psi = \sqrt{\frac{2}{a}}\sin\frac{n\pi x}{a}$$

where n and a are constants, is an eigenfunction of the Hamiltonian operator in one dimension:

$$\hat{\mathbf{H}} = -\frac{h^2}{8\pi^2 m}\frac{\partial^2}{\partial x^2}$$

when the potential energy is zero. What are the eigenvalues? Take m and h to be constant.

9. Show that the function $\phi = xe^{ax}$ is an eigenfunction of the operator:

$$\hat{\mathbf{O}} = \frac{d^2}{dx^2} - \frac{2a}{x}$$

where a is a constant. What are the eigenvalues?

10. Use the two-dimensional rotational operator

$$\begin{pmatrix} \cos\theta & -\sin\theta \\ \sin\theta & \cos\theta \end{pmatrix}$$

to find the new coordinates of the given point after rotation through the given angle:

(a) (2, 2) through 30°

(b) (4, 1) through 45°

(c) $(-4, -3)$ through 180°

(d) (3, 2) through 60°

(e) $(1, -3)$ through 240°

11. The BF_3 molecule is planar, with the fluorine atoms lying at the corners of an equilateral triangle and the boron atom at the center of the triangle. By assigning x-, and y-coordinates to each fluorine atom (the boron atom is placed at the origin), show that a two-dimensional $\mathbf{C}_3$ operation perpendicular to the plane of the triangle and through the boron atom will transform the molecule into itself. (*Hint*: Place one B–F bond along the y-axis.)

12. The differential operator for angular momentum is given by the expression

$$\hat{\mathbf{M}} = \frac{h}{2\pi i}(\mathbf{r} \times \nabla)$$

where h is a constant, $\mathbf{r} = \mathbf{i}x + \mathbf{j}y + \mathbf{k}z$, and

$$\nabla = \mathbf{i}\frac{\partial}{\partial x} + \mathbf{j}\frac{\partial}{\partial y} + \mathbf{k}\frac{\partial}{\partial z}$$

Assuming that $\hat{\mathbf{M}} = \mathbf{i}\hat{M}_x + \mathbf{j}\hat{M}_y + \mathbf{k}\hat{M}_z$, find the components $\hat{M}_x$, $\hat{M}_y$, and $\hat{M}_z$ of $\hat{\mathbf{M}}$.

13. Transform the components $\hat{M}_x$, $\hat{M}_y$, and $\hat{M}_z$ of angular momentum found in Problem 12 to spherical polar coordinates. (See Appendix III.)

14. Derive an expression for the total squared angular momentum operator

$$\hat{\mathbf{M}}^2 = \hat{M}_x^2 + \hat{M}_y^2 + \hat{M}_z^2$$

in spherical polar coordinates, using the operators found in Problem 13. (See Section 10-1 and Appendix III.)

CHAPTER 11 Numerical Methods and the Use of the Computer

11-1 INTRODUCTION

The computer has become an integral part of physical chemistry and, indeed, is a major part of our lives in general. In the second edition of this text, we included a chapter on computer programming, using a version of BASIC called FutureBASIC II for Macintosh computers.[1] Over the past few years, however, there has been a move away from using compiled programs for doing scientific computations and toward the use of spreadsheets, such as Microsoft Excel®, for such computations. While a simple spreadsheet cannot do some computations as well as a compiled program, it can handle most calculations associated with an undergraduate physical chemistry course. Moreover, spreadsheets are recognized for their ease in presenting data graphically, something that cannot be done easily with the output of a computer program. In addition, we shall see that when a macro (a computer program that can be called from within another program, such as a spreadsheet) is used along with the spreadsheet, the spreadsheet becomes a powerful tool indeed!

In this chapter, then, we shall concentrate on some of the applications of the spreadsheet to physical chemistry calculations, using Excel for the Macintosh. Versions of Excel for Windows-based computers are similar. We point out immediately, however, that the chapter is not intended to teach students how to use Excel with all its various features. It would take an entire text to do that, and many excellent texts are available on how to use Excel. Thus, we assume that students have at least an elementary understanding of the use of Excel. But, as with learning to program a computer, the best teacher is to do it. So, study the examples used in the various sections of the chapter and apply them to your problems. Also, for those students who feel that they need more background in learning Excel and applying it to chemistry, we recommend the text *How to Use Excel® in Analytical Chemistry and in General Scientific Data Analysis*, by Robert de Levie.[2] Many of the ideas presented in this chapter are drawn from this text.

[1]FUTUREBASIC II is a registered trademark of Staz Software, Inc., 3 Leisure Time Drive, Diamondhead, MS 39525-3215.
[2]New York: Cambridge University Press, 2001.

Before we begin, let us review some common mathematical operations and functions in Microsoft Excel. First, Excel distinguishes between text input, numerical input, and formula input. Text input in a cell is left justified, numerical input is right justified, and formula input must begin with an equals (=) sign. The three most commonly used keys in Excel are ENTER, TAB, and ESC. After typing text, numerical data, or formulas into an activated cell (an activated cell is surrounded by a heavy boundary), pressing ENTER will enter the information into the cell and then activate the cell immediately below it in the column (i.e., A1 → A2). Pressing TAB will enter the information into the activated cell and then activate the cell to the right in the next column (i.e., A1 → B1). Pressing ESC after typing information into a cell will delete the information from the cell.

Some common mathematical operations used in Excel are similar to those appearing in programming languages:

Operator	*Definition*
+	addition
−	subtraction
*	multiplication
/	division
^	exponentiation
pi()	$\pi = 3.14159\ldots$

Some common mathematical functions used in physical chemical calculations are as follows:

Function	*Returns*
AVERAGE(*range*)	average of numbers in a range (e.g., C3:C9)
COS(x)	cosine of angle in radians
EXP(x)	e^x
LN(x)	$\ln x$
LOG(x)	$\log_{10} x$
SIN(x)	sine of angle in radians
SQRT(x)	square root of x
SUM(*range*)	sum of numbers in a range (e.g., B1:B18)
TAN(x)	tangent of angle in radians

A computer should not be used as a calculator. Using a computer to do a single calculation such as $P = nRT/V$ does not make much sense, since it probably would take much less time to punch the data directly into a calculator than it would to set up the computer to do the calculation. It does make sense, however, to use the computer to calculate V in the van der Waals equation,

$$\left(P + \frac{n^2 a}{V^2}\right)(V - nb) = nRT \tag{11-1}$$

by successive approximations, or to calculate the Fourier transform of a wave, because these are time-consuming repetitive calculations on a calculator; the computer can do them in the blink of an eye. So we see that computer calculations are most useful when they involve iterative calculations and spreadsheets are especially useful when the output data are to be expressed graphically.

Example

As an introduction to computer methods using a spreadsheet, let us solve for V in the van der Waals equation, using the method of successive approximations. (This method also can be used to solve for n.) We shall approach the problem with a simple spreadsheet. The van der Waals equation is a cubic equation in V that has no analytical solution $V = f(T, P)$. To begin the method of successive approximations, we approximate a solution of the equation by dropping the n^2a/V^2 term from it and solving the equation for V. This gives

$$V = \frac{nRT}{P} + nb \tag{11-2a}$$

Once an approximate volume is found from Equation (11-2a), this volume is used as V in the n^2a/V^2 term in the equation

$$V = \frac{nRT}{\left(P + \frac{n^2 a}{V^2}\right)} + nb \tag{11-2b}$$

and a new V is calculated with this equation. This new volume is then used as V is the n^2a/V^2 term in Equation 11-2b, and another new volume is calculated. The process is repeated until there is no appreciable change in the newly calculated volume. In this example, let n = 50.0 moles, P = 20.0 bars, T = 300 K, a = 1.408 $\ell^2 \cdot \text{bar} \cdot \text{mol}^{-2}$, b = 0.03913 $\ell \cdot \text{mol}^{-1}$, and R = 0.08314 $\ell \cdot \text{bar} \cdot \text{mol}^{-1} \cdot \text{K}^{-1}$. Solving for V in Equation (11-2a), we have

$$V = \frac{(50.0)(0.08314)(300)}{(20.0)} + (0.03913)(50.0) = 64.31\ \ell$$

We now set up the spreadsheet:

1. Open Excel.
2. In cell A1, type V. Press **ENTER.**

3. In cell A3, type `64.31`. Press **ENTER**.
4. Calculate *nRT* using the preceding data, and in cell B1, type `nRT = 1247.1`. Press **TAB**.
5. Calculate n^2a using the preceding data, and in cell C1, type `n^2a = 3520`. Press **TAB**.
6. In cell D1, type `P = 20.0`. Press **TAB**.
7. Calculate *nb* using the preceding data, and in cell E1, type `nb = 1.957`.
8. In cell B3, type `=(1247.1/(20.0+(3520/(A3*A3))))+1.957`. Press **ENTER**. The first approximation should appear in cell B3. Make sure to match all the parentheses when you type in formulas.
9. In cell A4, type `= B3`.
10. Now click on cell B3, and drag your cursor down to cell B4, highlighting it.
11. Go up to the **EDIT** menu and scroll down to **FILL**. In the **FILL** box, move to, and click on, **DOWN**. The second approximation will appear in cell B4.
12. Now click on cell A4. Holding the mouse button down, move to cell B4 to highlight both cells A4 and B4, and then pull down (say, to about A12) to highlight several cells. In "Excelese," this operation generally is designated "Highlight A4:B12."
13. Go back to the **EDIT** menu, scroll down to **FILL**, and click on **DOWN**. The spreadsheet will recalculate *V*, using the value from the preceding row. After several iterations, the volume should be a constant. In this example, $V = 61.54\ \ell$.

Note that it took about six iterations to get the volume constant to two places past the decimal point. If the volume is not constant to the precision that you are after by the last iteration, repeat step 12, say, down to A24, etc. A representation of the Excel worksheet for this calculation is shown in Fig. 11-1.

	A	B	C	D	E
1	V	nRT=1.247.1	n^2a=3520	P=20.0	nb=1.957
2					
3	64.31	61.7667658			
4	61.7667658	61.5622848			
5	61.5622848	61.5447999			
6	61.5447999	61.5432972			
7	61.5432972	61.543168			
8	61.543168	61.5431569			
9	61.5431569	61.5431559			
10	61.5431559	61.5431559			
11	61.5431559	61.5431559			
12	61.5431559	61.5431559			
13					
14					

Figure 11-1 Excel worksheet for successive approximation calculation.

11-2 GRAPHICAL PRESENTATION

One of the most useful features of spreadsheets is their ability to express graphical data easily. Let us illustrate this feature with a graphical presentation of a sine wave.

Example

1. Open Excel.
2. In cell A1, type `t`. Press **TAB**.
3. In cell B1, type `f(t)=3sin(πt/8)`. Press **ENTER**.
4. In cell A2, type `0`. Press **ENTER**.
5. Click on cell A2 to highlight it. This will frame the cell in blue.
6. Go up to the **EDIT** menu and scroll down to **FILL**. Scroll down the **FILL** box to **SERIES**, and click on it.
7. In the **SERIES** box, check the following: Series in: **COLUMN**; Type **LINEAR**, **Step Value:** 1; **Stop Value:** 16. Click **OK**. This should list a series of numbers from 0 to 16 in steps of 1. The series corresponds to one period of the sine wave (i.e., from 0 to 2π).
8. In cell B2, type `=3*sin(pi()*A2/8)`. Click **ENTER**. Cell B2 shows sin (0).
9. Click on cell B2, and, while holding the mouse button down, pull down to highlight cells B2:B18.
10. Go to the **EDIT** menu, scroll down to **FILL**, and click on **DOWN**. (This procedure is used so often in spreadsheet calculations that from now on we shall refer to it simply as **FILL DOWN**).
11. The spreadsheet will list $f(t) = 3\sin(\pi t/8)$ for each value of t.
12. Click on cell A2, move the cursor to cell B2, and pull down to highlight all the cells with data (A2:B18).
13. Go up to the **INSERT** menu, scroll down to **CHART**, and click on it.
14. The Chart Wizard lists several types of standard charts. Highlight **XY SCATTER**. Do not choose the **LINE** type of chart, since it plots y versus category (the column number), rather than y versus x, spacing the categories equally along the x-axis. This sometimes can give erroneous graphs, particularly if the x-values are not equally spaced.
15. Choose the *type* of *XY SCATTER*. In this example, we shall choose data points connected by a smooth curve. Click on **NEXT**. This window shows the data range. Click on **NEXT**. This window allows one to label the chart and the various axes. Let us label the chart "Sine Function", the x-axis "time (seconds)", and the y-axis "3 $\sin(\pi t/8)$".
16. Click **FINISH**.

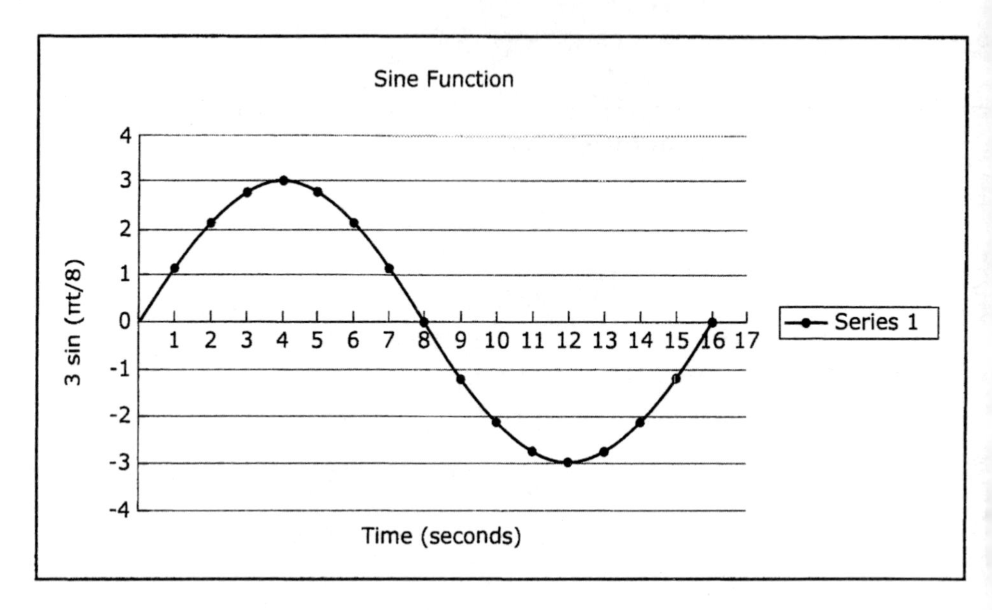

Figure 11-2 Spreadsheet graphical presentation of 3 $\sin(\pi t/8)$.

There are a number of things we can do with the graph after it is completed. For example, the colored background (blue in many cases) does not reproduce well. To get rid of the color, double-click on the chart area. A format window will open, allowing one to do all kinds of things to the area. For example, in the Area section, clicking on **NONE** will remove the colored background. Also, the curve itself may be in blue, which does not reproduce well. If so, double-click on the curve. A format window will open. In the **Patterns** folder, go to **Line** and then Color, pull down the color pallet, and click on black. Go back to the chart. The curve should appear in black. Now go back to the chart and double-click on the x-axis. A format window will open. Select the Scale folder and change Major Unit from 5 to 1. Note that now the time axis shows tick marks every second, instead of every 5 seconds. Play around with some of the other features. To insert the chart into a word-processing document, click on the chart border, which selects the chart. Go to the EDIT menu and click on COPY. Go to the word-processing document, and select an area in which the chart is to be placed. Go to the EDIT menu and click on PASTE. The chart should appear in the word-processing document. The Excel chart for 3 $\sin(\pi t/8)$ versus t is illustrated in Fig. 11-2.

Equations of Experimental Curves One of the best methods for obtaining physical constants from experimental curves is the *method of least squares.* We shall explore an actual least-squares calculation for linear equations in Chapter 12. Excel allows one to determine physical constants for experimental curves described by the following equations: linear, logarithmic, polynomial, power series, exponential, and moving average. To do this, Excel uses a routine called Trendline. Let us illustrate the use of this routine with a couple of physical chemistry examples.

Examples

1. A common experiment performed in many physical chemistry laboratory courses is the determination of partial molar volume, found by plotting the apparent molar volume of a solution against the square root of the molality. The result should be a straight-line graph. Part of the calculations for this experiment require the student to determine the y-intercept of the line. Consider the data taken from a series of solutions for NaCl in water, shown in Table 11-1. The apparent molar volume is defined by the equation

$$V_m = \frac{1}{m}\left(\frac{1000 + mM}{\rho} - \frac{1000}{\rho_o}\right)$$

where m is the molality of the solution, M is the molar mass of the solute, ρ is the density of the solution, and ρ_0 is the density of pure water.

Using the data from Table 11-1, let us set up the spreadsheet:

1. Open Excel.
2. In cell A1, type `m`. Press **TAB**.
3. In cell B1, type `m^1/2`. Press **TAB**.
4. In cell C1, type `V(m)`. Press **ENTER**.
5. In cells A2:A6, insert the data for m from Table 11-1.
6. In cell B2, type `=SQRT(A2)`.
7. Click on cell B2 and highlight cells B2:B6.
8. Go to **EDIT** and scroll down to **FILL → DOWN**. This will calculate the square root of the molalities.
9. In cells C2:C6, insert the data for V_m from Table 11-1.

TABLE 11-1 DATA FOR GRAPH OF APPARENT MOLAR VOLUME VERSUS SQUARE ROOT OF MOLALITY

m	V_m(ml)
3.2080	23.226
1.5625	21.195
0.7632	19.609
0.3816	18.793
0.1908	18.096

10. Highlight cells B2:C6.
11. Go to the **INSERT** menu and scroll down to **CHART** to bring up the Chart Wizard.
12. Scroll down to **XY SCATTER** and click on **Chart Sub-type 1: Scatter. Compares pairs of points.** This places the points on the graph without drawing the line.
13. Click **NEXT**. Click **NEXT** again. Label the graph and the coordinate axes.
14. Now, move up to the **CHART** menu and scroll down to **ADD TRENDLINE**.
15. Highlight the **Linear** box and click on **OPTIONS**.
16. Select "Display equation on chart."
17. Press **OK**. Trendline will place a heavy line through the points.
18. Double click on the heavy line. This opens the Format Trendline box. Under **PATTERNS**, click on **WEIGHT** and select the line of lighter weight. Press **OK**.

The completed chart is shown in Fig. 11-3. Note that the slope of the line is 3.803 and the y-intercept is 16.404.

The chart illustrated in Fig. 11-3 is shown as it is prepared by Excel in rough form (with the blue background removed). Let us now put the graph in a form more suitable for, say, a physical chemistry laboratory report:

1. First, it is not necessary to have the legend on the graph. Go to the **CHART** menu and scroll down to **Chart Options**. Click on the **Legend** folder, and remove the check from the "Show Legend" box.

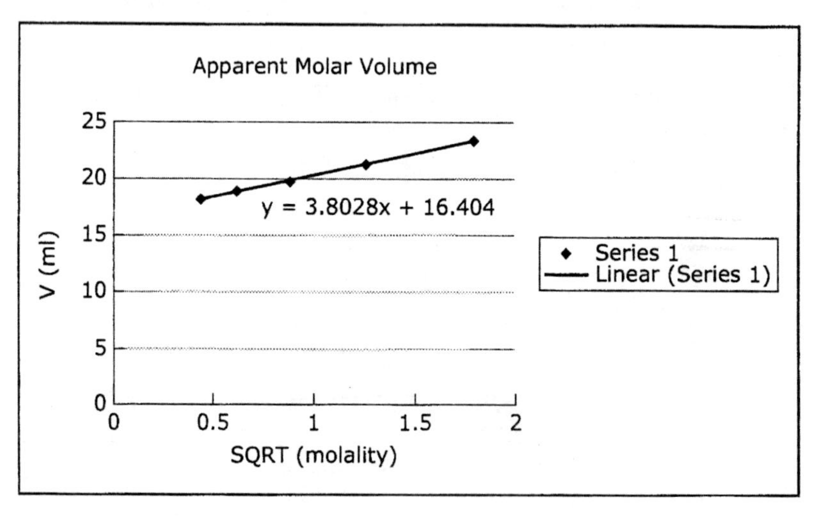

Figure 11-3 Graph of V_m versus $\sqrt{m}$ for NaCl solutions as prepared by Excel.

2. Double click on one of the Value(Y) Axis Major Grid lines that run across the graph area. In the **Patterns** folder, select "Color—Automatic" and change it to white. Press **OK**.
3. Double click on the y-axis (V-axis). The **Format Axis** box should appear. Go to **Scale**, and under **Minimum**, type 15. Press **OK**. This expands the scale and allows the data points to fill the graph area better.
4. Since the y-intercept is an important part of this experiment, it would be useful to have the line extend to $\sqrt{m} = 0$. The last data point on the graph is $\sqrt{m} = 0.4368$. Double click on the Trendline. The **Format Trendline** box will appear. Click on **OPTIONS**. In the **Forecast—Backward** box, type 0.4368. Press **OK**. The graph will extend back to the y-axis and show the y-intercept.
5. The equation of the line is expressed as $y = f(x)$. Let us change this to $V_m = f(\sqrt{m})$. Click on the equation for the line. Change `y` to `Vm` and `x` to `m^1/2`.

We see that there is a vast number of things that one can do to graphs to put them in a form suitable for scientific presentation. A graph more suitable to be used as part of a laboratory report is shown in Fig. 11-4.

2. Heat capacity is known to be a function of temperature. It is customary to express the heat capacity of a gas as the polynomial $C_P = a + bT + cT^2$. Table 11-2 lists heat capacity data for ammonia gas from 300 K to 750 K. Let us determine the polynomial constants a, b, and c from these data:

   1. Open Excel.
   2. In cell A1, type T

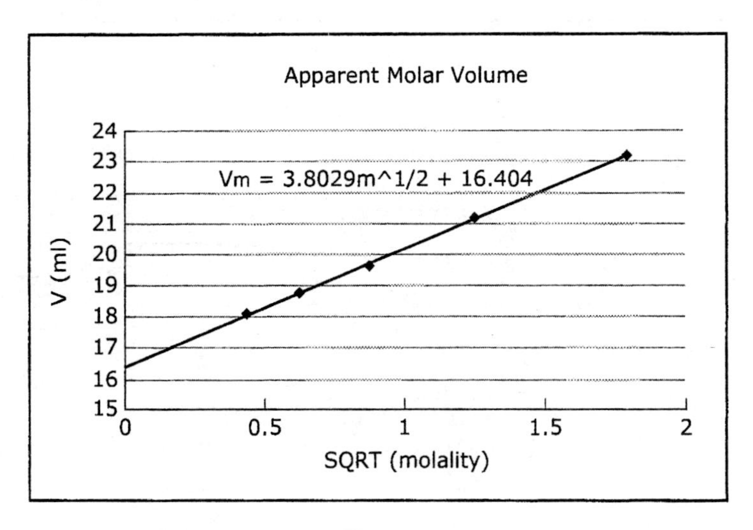

Figure 11-4 Graph of V_m versus $\sqrt{m}$ for NaCl solutions (modified from Excel).

TABLE 11-2 HEAT CAPACITY DATA FOR AMMONIA AS A FUNCTION OF TEMPERATURE

T (K)	C_P ($\mathrm{J \cdot mol^{-1} \cdot K^{-1}}$)	T (K)	C_P ($\mathrm{J \cdot mol^{-1} \cdot K^{-1}}$)
300	35.530	550	43.131
350	37.075	600	44.601
400	38.619	650	46.067
450	40.131	700	47.510
500	41.640	750	48.941

3. In cell B1, type C
4. In cells A2 to A11, insert the data for temperature from Table 11-2.
5. In cells B2 to B11, insert the data for C_P from Table 11-2.
6. Following the procedure found in the previous example, prepare an XY-Scatter graph for these data. Notice that the points appear almost to fall on a straight line.
7. Go to the **CHART** menu and scroll down to **ADD TRENDLINE.**
8. Highlight the Polynomial box. In this example, we want Order = 2 (second degree).
9. Open the **OPTIONS** folder and click on the "Display equation" box. Click **OK**.

The chart prepared here is illustrated in Fig. 11-5. Again, let us put the graph in a form more suitable for a laboratory or scientific report. First, let us

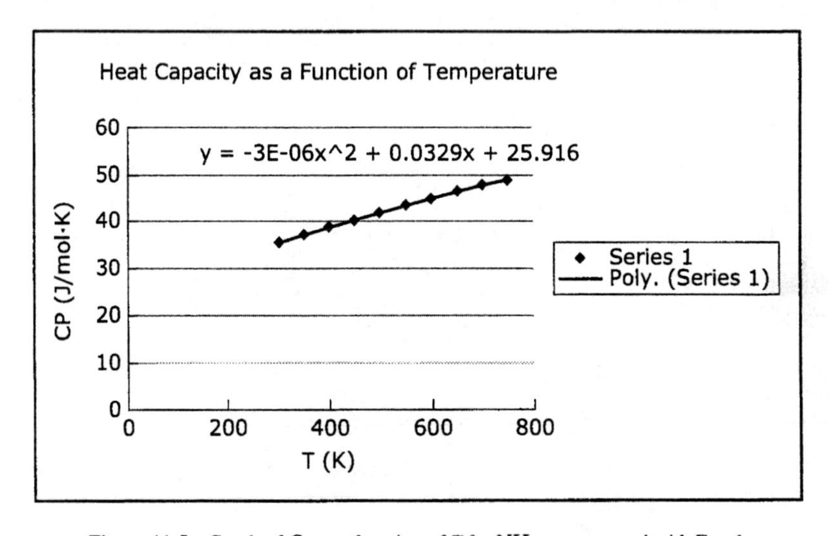

Figure 11-5 Graph of C_p as a function of T for NH_3 gas prepared with Excel.

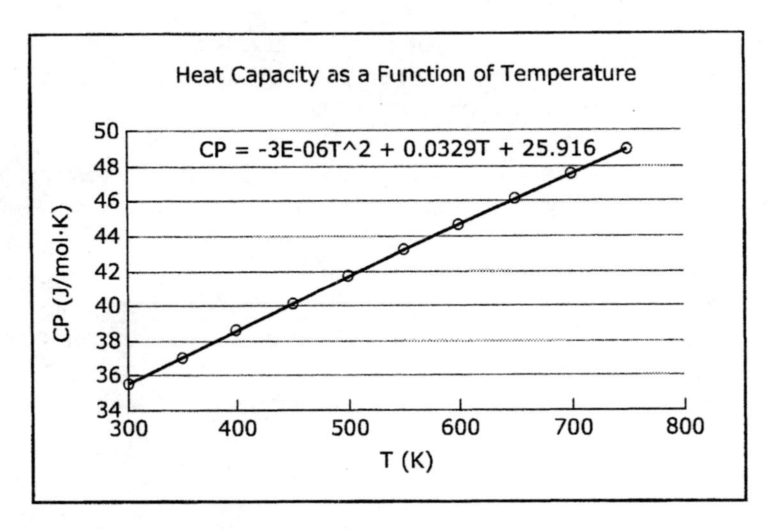

Figure 11-6 Graph of C_p as a function of T for NH_3 gas (modified from Excel).

change the style of the data points from diamonds to open circles. To do this, double-click on a point. This has to be done carefully, or you will click on the line. Bring the arrow slowly up to a point until the message box near the arrow indicates Series 1 and gives the coordinates of the point. Then double click on the point. This will open the **Format Data Series** box. Select the **PATTERNS** folder. In the **MARKER** section, choose a style (we chose circles), foreground (we chose black), and background (we chose no color). Click **OK**. This produces open circles. Make other modifications to the graph, as we did in Example 1. The completed graph is illustrated in Fig. 11-6. Note that the values for a, b, and c in the heat capacity equation are 25.916, 0.0329, and -3×10^{-6}, respectively. This agrees well with the published values of 25.90, 33.00×10^{-3}, and -30.4×10^{-7}.

11-3 NUMERICAL INTEGRATION

In Chapter 5, we defined the geometric interpretation of the definite integral as the area under the curve between the limits of integration. There are at least two situations in which analytical integration is not possible and yet it would be useful to know the area under the curve. The first is when the equation of the curve is not known. In these cases, some form of graphical integration (described in Chapter 12) can be performed. The second case is when the equation of the curve is known, but the function cannot be integrated analytically, such as the function $e^{\pm ax^2}$. In these cases, some form of numerical approximation to the area, called *quadrature*, can be used. In fact, we shall show that, even in the first case, numerical integration is useful.

A large number of numerical methods of integration have been developed, and many of them are particularly suited to the computer. In this section, we shall concentrate on two that are suited to spreadsheets.

Integration by the Trapezoid Method One of the simplest and most straightforward methods of integration involves dividing the interval between the limits of integration into a number of equal subdivisions and then extending vertical lines from the x-axis of the coordinate system to the curve. The points where these lines intersect the curve are connected together by straight lines forming a series of trapezoids, as shown in Fig. 11-7. The sum of the areas of the trapezoids closely approximates the area under the curve.

The area of a trapezoid is $A = \frac{1}{2}w(c + d)$, where w is the width of the base and c and d are the lengths of the sides. If we assume that $y = f(x)$ represents the equation of the curve, then the area of one of the trapezoids is

$$A = \frac{1}{2}\Delta x[f(x_1) + f(x_2)] \tag{11-3}$$

The sum of the areas of two adjacent trapezoids is

$$\begin{aligned} A &= \Delta x\left[\frac{f(x_1)}{2} + \frac{f(x_2)}{2} + \frac{f(x_2)}{2} + \frac{f(x_3)}{2}\right] \\ &= \Delta x\left[\frac{f(x_1)}{2} + f(x_2) + \frac{f(x_3)}{2}\right] \end{aligned} \tag{11-4}$$

We see, then, that when the curve is divided into n intervals, the total area under the curve between a and b is

$$A = \Delta x\left[\frac{f(a)}{2} + f(x_1) + f(x_2) + \cdots + f(x_{n-1}) + \frac{f(b)}{2}\right] \tag{11-5}$$

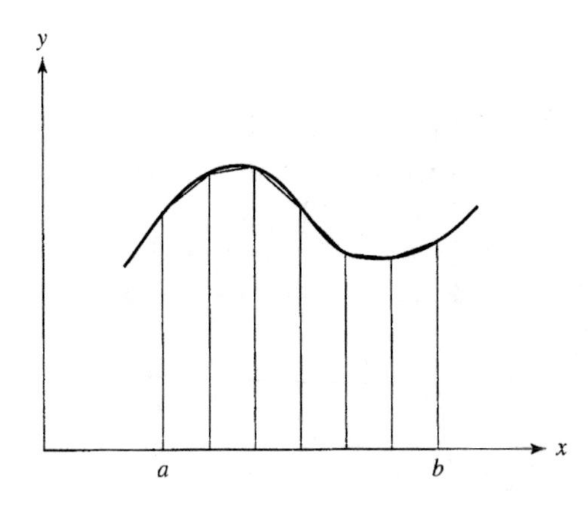

Figure 11-7 Integration using the trapezoid method.

Example

Let us now use Excel to determine the area under a curve by the trapezoid method. We choose a function that we can integrate analytically, so that we can compare our answer with the true integrated area. Determine the definite integral

$$y = \int_0^{\pi} \sin x \, dx$$

using the trapezoid method. We shall divide the area into 100 intervals. Therefore, Δx will be 0.0314159. Here are the Excel steps:

1. Open Excel.
2. In cell A1, type x.
3. In cell B1, type `sin x`.
4. In cell A3, place a zero.
5. Click on cell A3 to highlight it. Then go to **EDIT**, scroll down to **FILL → SERIES**, and click on it. In the Series Box, click on **COLUMNS**, then **Step value** = 0.0314159, and then **Stop value** = 3.1416. Click **OK**.
6. In cell B3, type `=sin(A3)/2`.
7. In cell B4, type `=sin(A4)`.
8. Click on cell B4 and highlight cells B4:B102. Go to **EDIT → FILL DOWN**.
9. In cell B103, type `=sin(A103)/2`.
10. In cell B104, type `=SUM(B3:B103)*0.0314159`.

The value in cell B104 is the answer, 1.9998, which is very close to the actual value, 2.0000.

Integration by the Newton–Cotes Method Numerical evaluation of the integral $\int_a^b f(x)\,dx$ requires replacing the integral with the sum

$$\sum A_n y_n = A_0 y_0 + A_1 y_1 + \cdots + A_n y_n$$

where the $n+1$ quantities of A_m are independent of the $n+1$ values of y_m. One common algorithm for finding A_m is based on the Newton–Cotes formula

$$A_m = \frac{(-1)^{n-m} h}{m!(n-m)!} \int_0^n \frac{z(z-1)(z-2)\cdots(z-n)}{(z-m)} dz \qquad (11\text{-}6)$$

where h is the common interval of x. Values for A_m calculated with Equation (11-6) are given in Table 11-3 for various values of n. Note that when $n = 1$, we have the

TABLE 11-3 NEWTON–COTES COEFFICIENTS FOR VARIOUS VALUES OF *n*

Order, n	Coefficients, A_m
1	$A_0 = A_1 = h/2$
2	$A_0 = A_2 = h/3;\ A_1 = 4h/3$
3	$A_0 = A_3 = 3h/8;\ A_1 = A_2 = 9h/8$
4	$A_0 = A_4 = 28h/90;\ A_1 = A_3 = 128h/90,\ A_2 = 48h/90$
5	$A_0 = A_5 = 95h/288;\ A_1 = A_4 = 375h/288;\ A_2 = A_3 = 250h/288$
6	$A_0 = A_6 = 246h/840;\ A_1 = A_5 = 1296h/840;\ A_2 = A_4 = 162h/840;\ A_3 = 1632h/840$

trapezoid method. When $n = 2$, we have another popular method, known as Simpson's rule. The advantage of using higher orders of the Newton–Cotes method becomes apparent when the functional dependence of y on x is not known and there may be very few data points in the interval between a and b. The only stipulation here is that the intervals along the x-axis be equally spaced.

Examples

1. The heat capacity at constant pressure of silver metal is given in Table 11-4 as a function of temperature from 20 K to 100 K. Let us evaluate the change in enthalpy of silver metal over this temperature range, using the Newton–Cotes method, given that

$$\Delta H = \int_{20}^{100} C_p\, dT$$

We can evaluate the integral by means of two five-point fits ($n = 4$): $T = 20$, 30, 40, 50, 60 and $T = 60, 70, 80, 90, 100$. The interval h in this case is 10. We use the coefficients for $n = 4$ given in Table 11-3. The Excel steps are as follows:

1. Open Excel.
2. In cell A1, type T.

TABLE 11-4 HEAT CAPACITY AT CONSTANT PRESSURE OF SILVER FROM 20 K TO 100 K

C_P (J · mol^{-1} · K^{-1})	T (K)	C_P (J · mol^{-1} · K^{-1})	T (K)
1.672	20	16.29	70
4.768	30	17.91	80
8.414	40	19.09	90
11.65	50	20.17	100
14.35	60		

3. In cell B1, type `Cp`.
4. In cells A3:A11, place the temperature data from Table 11-4.
5. In cells B3:B11, place the C_P data from Table 11-4.
6. In cell C7, type `=10*(28*B3+128*B4+48*B5+128*B6+28*B7)/90`.
7. In cell C8, type `=$C$7+10*(28*B4+128*B5+48*B6+128*B7+28*B8)/90`.
8. Click on cell C8 and highlight C8:C11. Go to **EDIT → FILL DOWN.**

The answer, 1034.3 J, is in cell C11.

2. Let us evaluate the integral $y = \int_0^{\pi} \sin x \, dx$, using only three 4-point fits $(n = 3)$. Thus, the interval between 0 and π is divided into nine parts. That is, $h = 3.14159/9 = 0.349066$. With the coefficients for $n = 3$ from Table 11-3, the following are the Excel steps:

   1. Open Excel.
   2. In cell A1, type `x`
   3. In cell B1, type `sin x`
   4. In cell A3, place a zero.
   5. Click on cell A3 to highlight it. Go to **EDIT → FILL → SERIES → COLUMNS → Step value:** 0.349066 **→ Stop value:** 3.1416.
   6. In cell B3, type `=sin(A3)`. Highlight cells B3:B12. Go to **EDIT → FILL DOWN.**
   7. In cell C6, type `=0.349066*(3*B3+9*B4+9*B5+3*B6)/8`.
   8. In cell C9, type `=C6+(0.349066*(3*B6+9*B7+9*B8+3*B9))/8`.
   9. Highlight cells C9:C12. Go to **EDIT → FILL DOWN.**

   The answer, 2.0004, is in cell C12. This example illustrates that, even with only nine equally spaced subintervals, the area can be very closely approximated.

11-4 ROOTS OF EQUATIONS

In Chapter 2, we described finding the roots of polynomial equations graphically. In this section, we shall discuss a method for determining the roots of polynomial equations in Excel with an operation known as *Goal Seek*. The algorithm for Goal Seek is based on the Newton–Raphson method, which, in turn, is based on the premise that a line drawn tangent to a curve described by $y = f(x)$ at the point x_1 will intersect the x-axis at a point x_2 having a value closer to the root than x_1. This situation is illustrated in Fig. 11-8. The slope of a line drawn tangent to a curve at a point x_1 is the derivative $y'(x_1) = dy/dx$. From the diagram, we see that

$$y'(x_1) = \tan\theta = \frac{y_1}{(x_2 - x_1)}$$

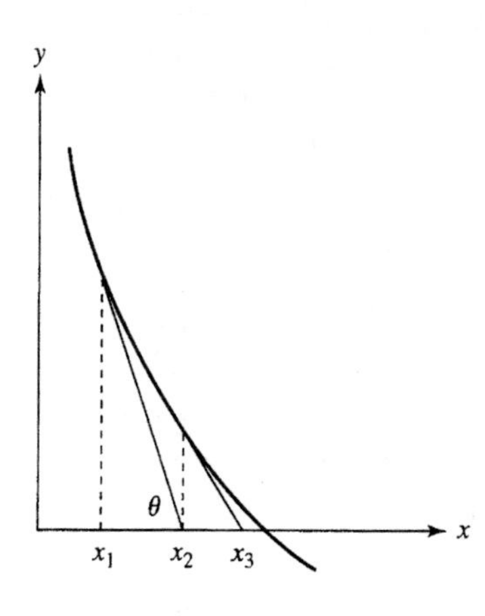

Figure 11-8 Illustration of the Newton-Raphson method.

or

$$x_2 = x_1 + \frac{y(x_1)}{y'(x_1)} \tag{11-7}$$

For the Newton–Raphson method to work, it is necessary that the process converge on the root within a reasonable number of iterations. This does not always happen, particularly if the initial guess is not close to the root. We find, however, that with Excel we can obtain a good initial guess that results in very close approximations to the roots of an equation.

Examples

1. Let us determine the roots of the equation $y = x^2 + 1.114x - 2.346$ with Excel. Since Goal Seek can do only one root at a time, we first need to determine the initial guess for each root. We shall do this with a not very elegant method, sometimes referred to as the brute-force approach:
   1. Open Excel.
   2. In cell A1, type `x`. (Remember to press **ENTER** or **TAB** after entering items into cells.)
   3. In cell B1, type `f(x)`.
   4. In cell A3, place a `-3`.
   5. Go to **EDIT** → **FILL** → **SERIES** → **COLUMNS** → **Step Value:** 0.1 → **Stop Value:** +3. Press **OK**.
   6. In cell B3, type `=A3^2+1.114*A3-2.346`.
   7. Click on cell B3 and highlight cells B3:B63. Go to **EDIT** → **FILL DOWN**.

8. Now go to the B column and look for the cells where the values change sign. Note the values for these cells in column A. In this example, A = −2.2 and A = 1.0. These are our initial guesses, since they are close to $f(x) = 0$.
9. Open a new workbook.
10. In cell B1, type `=A1^2+1.114*A1-2.346`.
11. In cell A1, place the first guess, `-2.2`. Press **ENTER**. Now click on cell B1 to highlight it.
12. Go to **TOOLS** → **Goal Seek**. The Goal Seek window should appear. **Set cell:** B1 → **To value:** 0 → **By changing cell:** A1. The Goal Seek Status window appears. Click **OK**. The value in cell A1 is the first root. In this example, $x = -2.187$.
13. In cell A2, place the second guess, `1.0`. Press **ENTER.** Go to cell B1, highlight B1:B2, and perform the **FILL DOWN** operation.
14. Click on cell B2 to highlight it. Go to **TOOLS** → **Goal Seek** → **Set cell:** B2 → **To value:** 0 → **By changing cell:** A2. The second root is 1.073.

2. In the Hückel molecular orbital treatment of the π-system of butadiene, one obtains a secular determinant having the characteristic equation $x^4 - 3x^2 + 1 = 0$. After obtaining our initial guesses, let us determine the roots of this equation using Goal Seek, which will give us the energy states:

1. Open Excel.
2. In cell A1, type `x`.
3. In cell B1, type `f(x)`.
4. In cell A3, place a `-3`. Use **EDIT** → **FILL** → **SERIES** to +3 with a step value of 0.1.
5. In cell B3, type `=A3^4-3*A3^2+1`. Use **FILL DOWN** to find the values of the equation for all values of x.
6. Find the x values where the equation changes sign, and place them in cells D3:D6. In this example, the x values are −1.7, −0.7, +0.6, and +1.6.
7. In cell E3, type `=D3^4-3*D3^2+1`. Use **FILL DOWN** to find the values of the equation for these values of x.
8. Click on cell E3 to highlight it. Use **Goal Seek** to find the value of D3 that will set E3 to zero.
9. Repeat Step 8 with E4, E5, and E6.
10. The roots are in cells D3, D4, D5, and D6.

The roots of the equation are $x = -1.618, -0.618, +0.618$, and $+1.618$.

One must be careful with multiple roots, such as double roots, and with imaginary roots: With these roots, when the "brute force" step is performed to obtain the initial guesses, the solution of the equation will never change sign. If the multiple roots are real, then the value of the equation will be zero at those points. If the roots are imaginary, then the equation will never reach the x-axis, much less cross it. A little mathematical intuition will be necessary here.

11-5 FOURIER TRANSFORMS REVISITED: MACROS

Digital computers do not handle continuous data very well. For this reason, the numerical calculation of Fourier transforms by computer requires discrete samples of the function.[3] For example, suppose we had some function of time, such as a sine wave or a cosine wave, expressed as $\sin(2\pi\nu t)$ or $\cos(2\pi\nu t)$, respectively. Since the function is periodic, we need sample it only over a single period τ_0, although in some cases sampling over more than one period could be useful. The discrete Fourier transform (DFT) of this function is defined as

$$F(\nu) = \frac{1}{N}\sum_k f(k)e^{2\pi i\nu k} \tag{11-8}$$

where N is the number of samples taken in the period τ_0 and k goes from $-t$ to $t - 1$ in integer steps centered around zero. The Fourier transform, then, is said to transform the function from a time domain to a frequency domain. Using Euler's relation $e^{i\theta} = \cos\theta + i\sin\theta$, we can write Equation (11-8) as

$$F(\nu) = \frac{1}{N}\sum_k f(k)(\cos 2\pi\nu k + i\sin 2\pi\nu k) \tag{11-9}$$

Thus, we can separate the Fourier transform into a real part and an imaginary part.

A major problem now is to determine how many times to sample the function over the period τ_0. Obviously, the more samples we take, the better the discrete transform will represent the continuous transform; but the more samples we take, the longer it will take to do the calculation. We find that computers can calculate discrete Fourier transforms very rapidly with the use of a type of algorithm called a fast Fourier transform (FFT). Such an algorithm requires that the number of sampling points be a power of two.

We find it convenient to calculate the real part and the imaginary part of Equation (11-9) separately. Let us define

$$\text{Re}(\nu) = \frac{1}{N}\sum_k f(k)\cos 2\pi\nu k \tag{11-10}$$

[3]Fourier transforms are discussed in Chapter 6.

and

$$i\,\mathrm{Im}(\nu) = \frac{i}{N}\sum_k f(k)\sin 2\pi\nu k \qquad (11\text{-}11)$$

The square of the absolute value of $F(\nu)$ is

$$|\mathrm{F}(\nu)|^2 = F^*(\nu)F(\nu) = (\mathrm{Re}(\nu))^2 + (\mathrm{Im}(\nu))^2 \qquad (11\text{-}12)$$

where $F^*(\nu)$ is the complex conjugate of $F(\nu)$. (See Chapter 1.) A graph of $|\mathrm{F}(\nu)|^2$ versus ν is called the *power spectrum* of the Fourier transform.

To illustrate how we use the preceding equations, let us calculate the discrete Fourier transform of the function $f(t) = 4\sin(\pi t/8)$. Let us use $N = 16$ sampling points over the period, which in this case is 16 seconds. Therefore, the sampling times will be from -8 seconds to $+7$ seconds in steps of 1 second. The corresponding sampling frequency for each sampling time is $\nu_k = t_k/16$. To find the power spectrum of the Fourier transform, we must first calculate $\mathrm{Re}(\nu)$ and $\mathrm{Im}(\nu)$ at the sampling frequencies. For example, to calculate $\mathrm{Re}(\nu)$ at $t_k = -8$ seconds ($\nu_k = -0.5$), we substitute $\nu = -0.5$ into Equation (11-10) and find the sum for $k = -8$ to $+7$. The pertinent data are listed in Table 11-5. We have

$$\begin{aligned}\mathrm{Re}(\nu = -0.5) = \frac{1}{16}[&0\cdot\cos(2\pi(-0.5)(-8)) - 1.531\cdot\cos(2\pi(-0.5)(-7))\\ &- 2.828\cdot\cos(2\pi(-0.5)(-6)) + \cdots\\ &+ 2.828\cdot\cos(2\pi(-0.5)(+6))\\ &+ 1.531\cdot\cos(2\pi(-0.5)(+7))] = 0\end{aligned}$$

This calculation is done for each frequency sampled. The results are given in Table 11-5. Note that the Fourier transform transformed the sine wave from a function of time to a single frequency $\nu_0 = 0.0625$ Hz and an amplitude. That is, $2\pi\nu_0 t = \pi t/8$ and therefore $\nu_0 = 1/16$. This is a long, tedious calculation even with only 16 sampling points. Let us see if we can get the computer to do the calculation. There is a way to introduce a program into a spreadsheet to make it perform desired computations and make logical decisions. Such programs are known as *macros* and can be called from within the spreadsheet. In Excel, macros are written in a dialect of BASIC called VBA (Visual BASIC for Applications). Since writing macros involves some understanding of programming in BASIC, we shall not deal with the actual writing of macros in this section. Rather, we shall concentrate on how to put a macro into an Excel spreadsheet, since there are many macros readily available from various sources, including the Internet.

TABLE 11-5 DISCRETE FOURIER TRANSFORM CALCULATION OF 4 sin($\pi t/8$)

Time (t_k)	$f(t_k)$	ν_k	Re(ν_k)	Im(ν_k)	$\lvert F(\nu_k)\rvert^2$
−8	0	−0.500	0	0	0
−7	−1.531	−0.438	0	0	0
−6	−2.828	−0.375	0	0	0
−5	−3.696	−0.313	0	0	0
−4	−4.000	−0.250	0	0	0
−3	−3.696	−0.188	0	0	0
−2	−2.828	−0.125	0	0	0
−1	−1.531	−0.0625	0	−2.000	4.000
0	0	0	0	0	0
1	1.531	0.0625	0	2.000	4.000
2	2.828	0.125	0	0	0
3	3.696	0.188	0	0	0
4	4.000	0.250	0	0	0
5	3.696	0.313	0	0	0
6	2.828	0.375	0	0	0
7	1.531	0.438	0	0	0

To illustrate how to put a macro into a spreadsheet, consider the following example taken from Robert de Levie's text *How to Use Excel in Analytical Chemistry and in General Scientific Data Analysis*, cited earlier in the chapter.[4]

```
'Macro to read a cell value
Sub read()
myValue = Selection.Value
MsgBox "The cell value is " & myValue
End Sub
```

Before entering this macro into a spreadsheet, let us look as some of its features, since they are common to all macros and important to our purpose here. The first statement, beginning with an apostrophe, is known as a comment statement. Comment statements are placed in a macro to benefit the user. They are ignored by the computer. One can have as many comment statements as one wishes in a macro, as long each comment line begins with an apostrophe. In Excel, comment statements generally appear in green. The second line designates the macro as a subroutine. We call it "read" in this example. The name of the macro always ends in (). In the next line, Selection. Value indicates that a particular cell in the spreadsheet is to be selected and that the spreadsheet will assign the variable myValue the numerical value placed in that cell. Note that the period between "Selection" and "Value" is important; do not forget to type it in. MsgBox stands for Message Box. The message in the Message Box is given in quotes. The symbol "&" informs the computer that the message is completed.

[4]Reprinted with the permission of Cambridge University Press.

Let us now place this macro into a spreadsheet:

1. Open Excel.
2. Go to **TOOLS → MACRO → Visual Basic Editor**. A Project—VBA Project window will open up. Click on Sheet 1. Press **RETURN**.
3. The Code window should open. Type the macro code for "read" exactly as it is shown in the Code window.
4. Go back to the spreadsheet. This can be done by clicking on it (usually you can see the spreadsheet in back of the Code window) or by pressing the keys Opt+F11. The keys Opt+F11 will allow you to toggle back and forth between the spreadsheet and the macro code window. Place several different numbers in various cells on the spreadsheet. Now click on a particular cell containing a number, in order to highlight the number. Go to **TOOLS → MACRO → Macros. . . .** The Macro window should open. Click on "Read" if it is not highlighted, and then press **RUN**. The computer will return to the spreadsheet, and a message box will open specifying the value in the chosen cell. The Message Box must be acknowledged in order to continue using the spreadsheet. Press **OK**. Try this with other cells.

There is a way to assign a control key to a particular macro, so that one does not have to keep going through the **TOOLS → MACRO → Macros. . . → RUN** procedure to run the macro:

5. While in the spreadsheet, go to **TOOLS → MACRO → Macros. . . .** In the Macro window, go to **OPTIONS**, and where it indicates **Shortcut key**, place an r in the box. (Any letter will do.)
6. Go back to the spreadsheet. Highlight a cell containing a number. Press Opt+Cmd+r. A message box will appear specifying the value in the chosen cell.

Now that we know how to introduce a macro into a spreadsheet, let us prepare a spreadsheet that will allow us to calculate discrete Fourier transforms:

1. Open Excel.
2. Go to **TOOLS → MACRO → Visual Basic Editor**. In the project box, click on **This workbook**. Press **RETURN**.
3. In the Code window, carefully type in the following little no-frills macro for calculating a discrete Fourier transform:

```
'Macro for calculating a discrete Fourier transform
Sub Fourier()
Dim dataArray As Variant, cn As Integer, cnMax As Integer,
  rn As Integer
Dim rnMax As Integer, i As Integer, j As Integer, nu(1000)
  As Double
Dim time(1000) As Double, fn(1000) As Double, Re(1000) As
  Double
```

```
Dim Im(1000) As Double, Pw(1000) As Double, A As Double, B
  As Double
dataArray = Selection.Value
rnMax = Selection.Rows.Count
cnMax = Selection.Columns.Count
For rn = 1 to rnMax
    nu(rn) = dataArray(rn, 3)
    fn(rn) = dataArray(rn, 2)
    time(rn) = dataArray(rn, 1)
Next rn
For j = 1 to rnMax
    A = 0
    B = 0
    For i = 1 to rnMax
        A = A + (fn(i) * cos(2 * 3.14159 * nu(j) * time(i)))
        B = B + (fn(i) * sin(2 * 3.14159 * nu(j) * time(i)))
    Next i
    Re(j) = A/rnMax
    Im(j) = B/rnMax
    Pw(j) = Re(j) * Re(j) + Im(j) * Im(j)
Next j
'Output data
For i = 1 to rnMax
    dataArray(i, 5) = Re(i)
    dataArray(i, 6) = Im(i)
    dataArray(i, 7) = Pw(i)
Next i
Selection.Value = dataArray
End Sub
```

Save this spreadsheet as Fourier Transform. Note that in the line A = A + (fn(i)..., both fn and time are indexed with i, but nu is indexed with j. The same is true for the next line, B = B + (fn(i).... Be aware of this distinction when you are typing in the macro.

To use the macro, the spreadsheet must be set up in a special way. Let us demonstrate this with several examples.

Examples

1. Calculate the power spectrum of the discrete Fourier transform of $4\sin(\pi t/8)$.
    1. Open the spreadsheet Fourier Transform.
    2. In cell A1, type `time`.
    3. In cell B1, type `f(t)`.
    4. In cell C1, type `nu`.
    5. In cell E1, type `Re`.

6. In cell F1, type `Im`.
7. In cell G1, type `Pw`.
8. In cell A3, place a `-8`. Using the **Series** command in menu **EDIT** → **FILL**, construct the series **Column** to +7 in steps of 1.
9. In cell B3, type `=4*sin(pi()*A3/8)`. Click on cell B3 and highlight down to all values in column A. Go to **EDIT** and **FILL DOWN**.
10. In this example a period, is 16 seconds. Therefore, in cell C3, type `=A3/16`. Click on cell C3 and highlight down to all values in column A. **FILL DOWN**.
11. This next step is critical, or the macro will not work. You must now click on cell A3 and highlight all the cells A3:G18 (i.e., all the cells that have or will have data).
12. Go to **TOOLS** → **MACRO** → **Macros . . .**. The macro Fourier should be highlighted. Press **RUN.** Fourier transform data appear immediately in columns E, F, and G and should match the data in Table 11-5.
13. Go to **TOOLS** → **MACRO** → **Macros . . .**. Select **OPTIONS**. Place an `f` in the **Shortcut Key** box.

If you forget to highlight the required columns, you will get an Error Box message that will say "Run-time error". If this happens, press **End** and repeat the macro, this time highlighting all the required cells.

If you wish to prepare a graph of the power spectrum, perform the following steps:

14. In cell D3, type `=G3`. Click on cell D3 and highlight cells D3:D18. Use **FILL** → **DOWN**.
15. Now highlight cells C3:D18. Go to **INSERT** → **Chart** → **XY Scatter**. Follow the procedure outlined in section 11-2 for preparing graphs.

In the next two examples, we shall show how a Fourier transform calculation can find the frequencies making up a complex wave. The ability to do this is the basis of Fourier transform infrared spectroscopy and Fourier transform nuclear magnetic resonance spectroscopy.

2. Find the Fourier transform of the function $f(t) = 2\cos(\pi t/8) + 3\cos(4\pi t/8)$. Prepare graphs of both $f(t)$ and the power spectrum of the Fourier transform.

   1. Open Fourier Transform.
   2. In cell A3, type `-8`. Press **RETURN**.
   3. Highlight cell A3 and go to **EDIT** → **FILL** → **Series** → **Column** → **Step value:** 0.5 → **Stop value:** 7.5. Press **OK**.
   4. In cell B3, type `=2*cos(pi()*A3/8)+3*cos(4*pi()*A3/8)`. Press **RETURN**.

5. Highlight cells B3:B34. Go to **EDIT → FILL DOWN.**
6. In cell C3, type `=A3/16`. Press **RETURN**. Highlight cells C3:C34. Go to **EDIT → FILL DOWN**.
7. Highlight cells A3:G34. Activate the macro Fourier.
8. In cell D3, type `=G3`. Press **RETURN**. Highlight cells G3:G34. **FILL DOWN**.
9. Prepare a graph of columns A3:B34 and a graph of columns C3:D34.

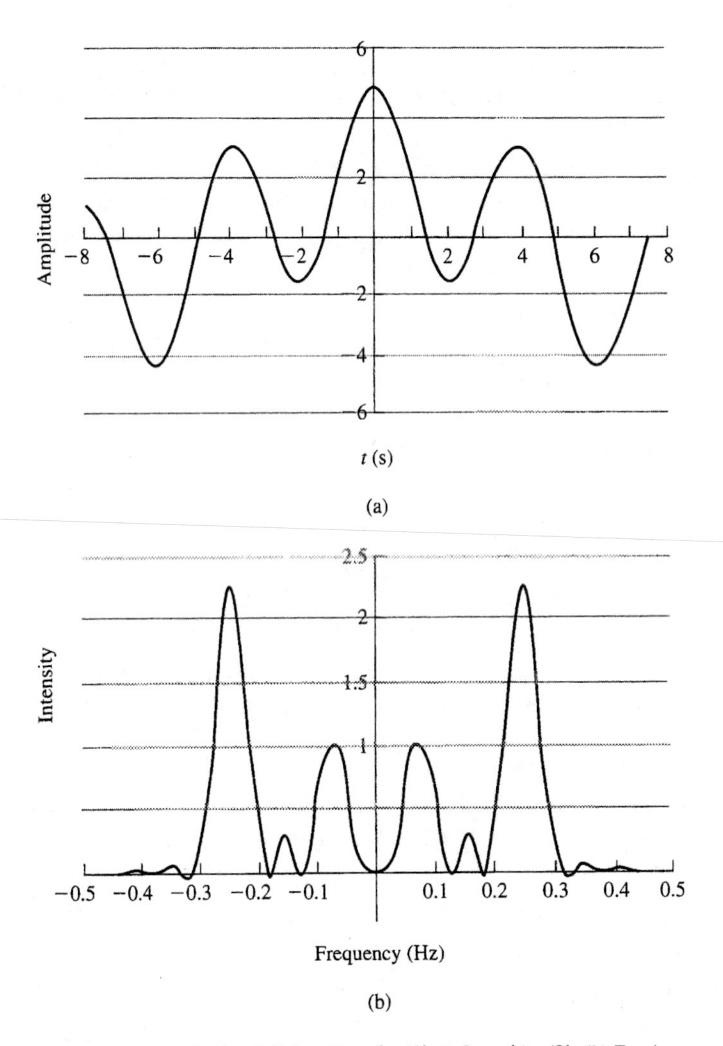

Figure 11-9 (a) Graph of $f(t) = 2\cos(\pi t/8) + 3\cos(4\pi t/8)$. (b) Fourier transform of $f(t)$.

Note that the Fourier transform gives the two frequencies, 0.0625 Hz and 0.250 Hz, and their respective amplitudes. Remember that the intensity of a wave, which Pw represents, is the absolute square of the amplitude. Also, the Fourier transform splits the amplitude into a positive and negative component. Therefore, the amplitude of the wave at 0.0625 Hz is $\sqrt{1.000} \times 2 = 2.000$, and the amplitude of the wave at 0.250 Hz is $\sqrt{2.25} \times 2 = 1.500 \times 2 = 3.000$. Graphs of the wave and its Fourier transform are shown in Fig. 11-9.

3. As we have mentioned, Fourier transforms have an important application in various forms of spectroscopy. Figure 11-10 (a) shows the nuclear magnetic resonance free-induction decay signal (FID) for a compound, such as

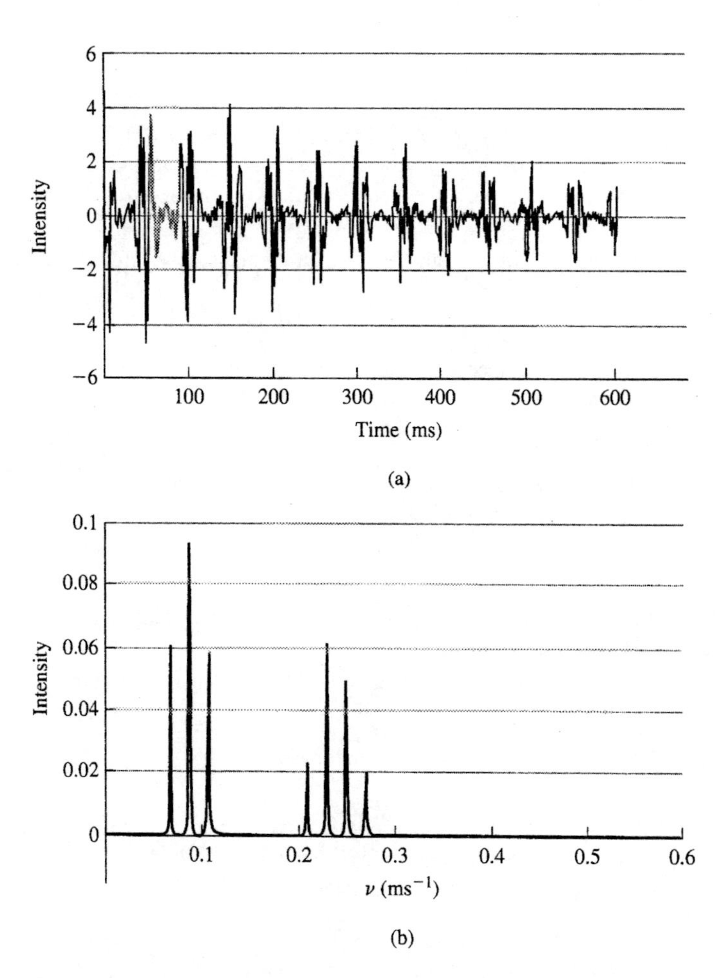

Figure 11-10 (a) Free-induction decay signal of CH_3CH_2X. (b) Fourier transform of FID signal.

CH_3CH_2Cl. Calculating the Fourier transform of this signal produces the spectrum shown in Fig. 11-10 (b). (See Problem 11-13.)

SUGGESTED READINGS

1. DE LEVIE, ROBERT. *How to Use Excel® in Analytical Chemistry and in General Scientific Data Analysis*. New York: Cambridge University Press, 2001.
2. DICKSON, T. R. *The Computer and Chemistry*. San Francisco: W. H. Freeman and Co., 1968.
3. NOGGLE, JOSEPH H. *Physical Chemistry on a Microcomputer*. Boston: Little, Brown, and Co., 1985.

PROBLEMS

1. A tank has a volume of 30.0 liters and holds NO gas at a pressure of 175.0 bars and a temperature of 27.0°C. Find the mass in grams of the NO in the tank, using the method of successive approximations with the van der Waals equation. In the equation, take $a = 1.358\ \ell^2 \cdot \text{bar} \cdot \text{mol}^{-1}$ and $b = 0.02789\ \ell \cdot \text{mol}^{-1}$.
2. The total volume of a solution is found to vary with molality m according to the equation

$$V = a + bm + cm^2$$

where a, b, and c are experimentally determined constants. The following data were collected for a sucrose–water solution containing 1000 g of water:

m	$V(cc)$	m	$V(cc)$
1.5000	1156.77	0.1875	1069.88
0.7500	1125.00	0.0938	1058.10
0.3750	1091.23		

Use Trendline to find the coefficients a, b, and c in the given equation by fitting the curve to a polynomial.

3. A set of kinetics data for a reaction is found to obey the Arrhenius equation

$$\ln k = \frac{-E_a}{RT} + \ln A_0$$

Given the following set of data for a first-order rate constant as a function of temperature, find the activation energy for the reaction, E_a, by plotting $\ln k$ versus $1/T$ and using the method of least squares to find the slope of the line:

$k(s^{-1})$	$T(K)$	$k(s^{-1})$	$T(K)$
1.780×10^{-4}	288.2	1.737×10^{-3}	308.2
5.777×10^{-4}	298.2	4.876×10^{-3}	318.2

4. Integrate the given integrals, using the trapezoid method. Express your answers to at least four significant figures.

(a) $\int_0^4 \sqrt{x+1}\, dx$ **(c)** $\int_0^{\pi/2} \cos x\, dx$

(b) $\int_1^4 \frac{1}{x} dx$ **(d)** $\int_1^2 e^{-x}\, dx$

5. The following table gives the molar heat capacity of $Al_2O_3(s)$ at temperatures from 30 K to 110 K:

C_P (J · mol^{-1} · K^{-1})	T (K)	C_P (J · mol^{-1} · K^{-1})	T (K)
0.263	30	6.895	80
0.691	40	9.69	90
1.492	50	12.84	100
2.779	60	16.32	110
4.582	70		

Find the change in enthalpy for $Al_2O_3(s)$ from 30 K to 110 K by the Newton–Cotes method, using two five-point fits $(n = 4)$. Remember that

$$\Delta H = \int_{30}^{110} C_P\, dT$$

6. The following table gives the molar heat capacity of platinum metal at temperatures from 5 K to 30 K:

C_P (J · mol^{-1} · K^{-1})	T (K)	C_P (J · mol^{-1} · K^{-1})	T (K)
0.1014	5	1.444	20
0.2185	10	2.673	25
0.6438	15	4.136	30

Find the change in the entropy of platinum metal from 5 K to 30 K by the Newton–Cotes method, using one six-point fit $(n = 5)$. Compare your answer with that using the trapezoid method. Remember that

$$\Delta S = \int_5^{30} \frac{C_P}{T} dT$$

7. From the data given in a paper by J. O. Hutchins, A. G. Cole, and J. W. Stout, *J. Amer. Chem. Soc.*, **82**, 4813 (1960), determine S^0 at 298 K for crystalline ℓ-alanine by the trapezoid method of numerical integration. Remember that

$$S^0 = \int_0^{298} \frac{C_P}{T} dT$$

8. Evaluate the integral

$$\frac{2}{a}\int_0^a \sin^2\frac{\pi x}{a}\,dx$$

by the Newton–Cotes method, using two four-point fits ($n = 3$). (*Hint*: Allow x to vary in units of a from 0 to 1.)

9. Find the roots of the following equations:

(a) $x^3 + 1.591x^2 - 4.464x - 1.631 = 0$

(b) $x^3 - 3x^2 + 4 = 0$

(c) $x^3 - 1.470x^2 - 0.654x + 1.170 = 0$

(d) $x^4 - 1.329x^3 - 5.651x^2 + 11.289x - 5.309 = 0$

10. The Hückel treatment of π system of vinyl chloride, CH_2CHCl, yields a secular determinant having the characteristic equation $x^3 + 2x^2 - 1.111x - 2 = 0$. Find the roots of this equation with Excel. Assume that the solutions lie within the range from -3 to $+3$.

11. Using the macro *Fourier*, find the discrete Fourier transform of the step function,

$$f(t) = \begin{cases} -1; & -\pi \le t < 0 \\ +1; & 0 < t \le +\pi \end{cases}$$

(*Hint*: Let t go from -0.50 to 0 in steps of 0.01563. Then type in a 0. Then let t go from 0 to $+0.4845$ in steps of 0.01563. This should result in $2^6 = 64$ sampling times. Let $\nu = 20t$. Let $f(t) = 0$ at $t = -0.5$ and $t = 0$, $f(t) = -1$ from $-0.5 < t < 0$, and $f(t) = +1$ from $0 < t < +0.5$.) Verify that the imaginary part of the transform leads to the eight-term series

$$\begin{aligned} f(t) = {} & 2[0.6895\sin(2\pi t) + 0.2001\sin(6\pi t) + 0.1248\sin(10\pi t) \\ & + 0.08628\sin(14\pi t) + 0.06811\sin(18\pi t)] \end{aligned}$$

12. Using Excel, prepare a graph of the series given in Problem 11 from $t = -0.50$ to $+0.50$ in steps of 0.01 and show that it follows the step function closely.

13. The free-induction decay signal shown in Fig. 11-10(a) can be simulated with the equation

$$\begin{aligned} f(t) = {} & e^{-t/500}[0.866\cos(2\pi(0.108)t) + 1.23\cos(2\pi(0.088)t) \\ & + 0.866\cos(2\pi(0.068)t) + 0.5\cos(2\pi(0.270)t) \\ & + 0.866\cos(2\pi(0.250)t) + 0.866\cos(2\pi(0.230)t) \\ & + 0.5\cos(2\pi(0.210)t)] \end{aligned}$$

where t is in milliseconds. Find the power spectrum of the Fourier transform of $f(t)$, allowing t to vary from 0 to 512 ms in steps of 1 ms and ν to vary from 0 to 0.512 ms^{-1} in steps of 0.001 ms^{-1}.

CHAPTER 12

Mathematical Methods in the Laboratory

12-1 INTRODUCTION

Most laboratory experiments in physical chemistry are concerned with measurements. There is a basic difference, however, between experiments performed at the undergraduate level and those performed in physical chemistry research laboratories. At the undergraduate level, students perform experiments that have a known outcome. Generally, these experiments have been performed many times over a number of years by numerous students. In research laboratories, by contrast, scientists usually perform experiments on unknowns. There are no laboratory instructors from whom a research scientist can obtain the correct answer to an experimental measurement to see whether he or she has performed the experiment correctly. Thus, it is important for students of physical chemistry, who hope someday to become proficient researchers, to learn how to assess the reliability of their experimental data. One common way of doing so is to perform the experiment more than once. It is known that when a measurement is made more than once, the results generally scatter around some average value. We shall see in the next few sections that this experimental scatter can be used to help estimate the probability that the average value is the "true" value. But before we can do that, let us review some simple probability theory.

12-2 PROBABILITY

The *probability* of any item having a specific characteristic is the number of items having that characteristic, divided by the total number of items in the assembly. For example, if we had a barrel containing 50 apples, 40 of which were red and 10 green, the probability of picking a green apple from the barrel (assuming an even distribution of the apples throughout the barrel) is

$$p = \frac{10}{50} = 0.20 \quad \text{or} \quad 20\%$$

TABLE 12-1 MASS OF BAR OF ALUMINUM METAL

Measured Value, m_i (g)	Frequency, n_i	Relative Frequency, n_i/N
8.81	2	0.04
8.82	2	0.04
8.83	11	0.22
8.84	17	0.34
8.85	14	0.28
8.86	3	0.06
8.87	1	0.02

Let us relate this definition of probability to a hypothetical collection of measurements. Suppose that we measure the mass of a block of aluminum on a beam balance and that we perform this measurement over and over, for a total of 50 measurements. In this example, we assume that the actual mass of the metal does not change (i.e., pieces of the metal do not break off, nor does the sample get dirty from handling); we assume also, to illustrate a point, that the measurements are spread over a much larger range than we would most probably see if we used a good laboratory balance. The results of the 50 measurements are listed in Table 12-1. The data also are plotted as bar graphs in Fig. 12-1. Note that the general shape of each bar graph does not depend on whether n_i or n_i/N is plotted against m_i. We find, however, that the height of the bars does depend on the difference between the measured values Δm. For example, if we had measured the mass to the nearest 0.005 gram

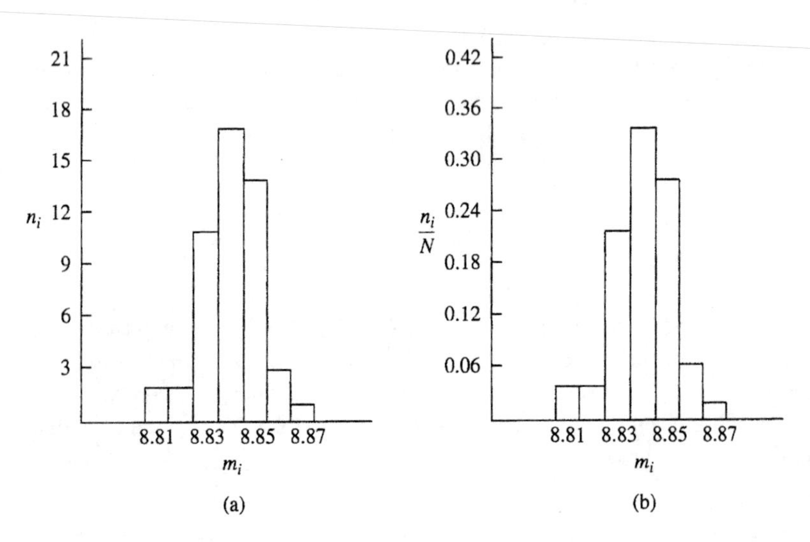

Figure 12-1 Distribution of measurements: (a) frequency versus measured values; (b) relative frequency versus measured values.

rather than to the nearest 0.01 gram, the 50 measurements would have been distributed over 13 data points (from 8.810 to 8.870 in steps of 0.005), rather than over the 7 data points shown, causing the height of each bar to be reduced.

To make the graphs more comparable, then, let us plot $n_i/(\Delta m \cdot N)$, rather than n_i/N, against m_i. Such a graph is illustrated in Fig. 12-2. Since the shape of this graph no longer depends on Δm, let us allow Δm to approach zero and greatly increase the number of measurements, N. Hence, we can write

$$\lim_{\Delta m \to 0} \frac{1}{N} \frac{n_i}{\Delta m} = \frac{1}{N} \frac{dn_i}{dm} = P(m) \tag{12-1}$$

where dn_i is the number of measured values lying between m_i and $m_i + dm$. The function $P(m)$ is represented by the dashed line in the figure. According to our definition of probability, the probability that any measurement will have a value lying between m_i and $m_i + dm$ is equal to the number of measurements lying in the range between m_i and $m_i + dm$, which is dn_i, divided by the total number of measurements N:

$$\text{Probability} = \frac{dn_i}{N} = \left(\frac{1}{N} \frac{dn_i}{dm}\right) dm = P(m)dm \tag{12-2}$$

The function $P(m)$ is called the *error probability function* and represents the distribution of the probability over the measured values. Such a distribution is called a *distribution aggregate*. Every measured value is an *element* of that aggregate.

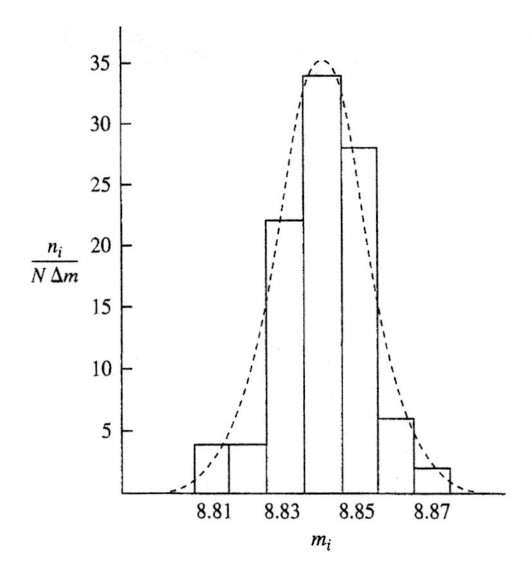

Figure 12-2 Distribution of measured values showing the error probability function $P(m)$.

12-3 EXPERIMENTAL ERRORS

Let us now relate what we found in the previous section to experimental errors. As we said earlier, when a measurement is made more than once, the results scatter about an average, or mean, value, which we can define as

$$\overline{m} = \frac{1}{N}\sum m_i \tag{12-3}$$

This experimental scatter is caused by a type of error called *random error* and is due to the fact that various measuring devices have inherent limits in the *precision*[1] to which they can be read. A second type of error, called *systematic error*, due to such things as uncalibrated instrumentation, human reaction times, and so on, generally is unidirectional. That is, systematic errors normally do not scatter around a mean value. We generally speak of systematic error as being related to experimental design. Therefore, while random error may help us determine the reliability of our data, performing a poorly designed experiment more than once will *never* help us get to the "true" answer.

While it is possible to discuss the random errors associated with measurements, we cannot determine what they actually are. We can, however, consider how far each measured value is from the mean value. We define this deviation from the mean, called a *residual*, as

$$r_i = m_i - \overline{m} \tag{12-4}$$

If the errors incurred in a particular experiment are only random errors and a large number of measurements is taken, the measurements fall on a normal distribution curve, such as the one described in the previous section. We will show that, under these conditions, the mean value is the *best value* with a higher precision than any single value. If the errors that are encountered are systematic errors, no statistical treatment of data can compensate for them. When the errors are entirely random, the error probability function, plotted as a function of the residual r, rather than the measured value, has the form

$$P(r) = \frac{1}{\sigma\sqrt{2\pi}}e^{-r^2/2\sigma^2} \tag{12-5}$$

[1]The precision of a measurement is the relative error between that measurement and other, similar measurements performed with the same measuring device. Precision should be contrasted with the *accuracy* of a measurement, which is the relative error between the measured value and the "true" value of the quantity being measured. Note that a set of highly precise measurements is not necessarily accurate. For example, if a student were to measure the temperature of a constant-temperature bath several times, using a thermometer that had not been calibrated, the results could be quite precise, but, since the thermometer was not calibrated, highly inaccurate.

where σ is a parameter known as the *standard error* or *estimated standard deviation.* This type of distribution is known as a *Gaussian distribution.* In terms of residuals, the standard error has the form

$$\sigma = \sqrt{\frac{1}{N-1}\sum r_i^2} \tag{12-6}$$

Note that the probability that a single measurement will have a deviation from the mean lying in the range between r and $r + dr$ is $P(r)dr$ and that this is just the area of the small shaded "rectangle" (it looks more like a trapezoid, because dr is so large in the diagram) of width dr, in Fig. 12-3. Since the total area under the probability distribution curve must represent the probability of finding the error lying between $-\infty$ and $+\infty$, and this must be 100%, or 1, we can write

$$\int_{-\infty}^{+\infty} P(r)dr = 1 \tag{12-7}$$

Equation (12-7) is called the *normalization equation,* since it guarantees that the sum of all the probabilities must equal 100%. If we assume that the error in a single measurement and the residual are essentially the same, then the probability that a measurement will have an error no larger than $\pm\sigma$ associated with it is found to be 68.3% of the total area under the probability curve (illustrated in Fig. 12-3).

Of the many types of parameters used to express uncertainty due to random error, the *probable error* is one of the most popular. The probable error, q, is defined as having a value such that the probability that an error in a measurement chosen at

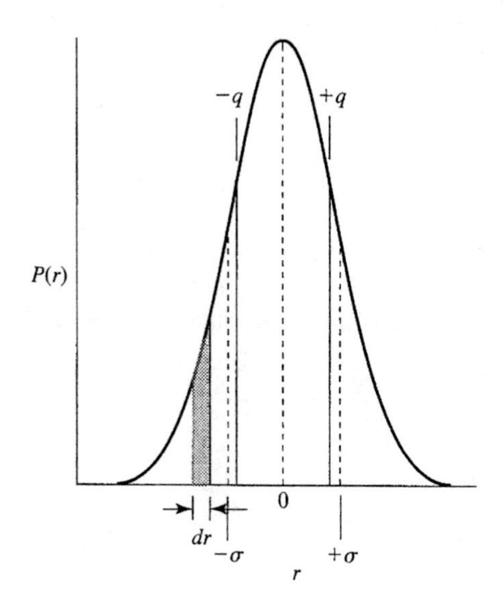

Figure 12-3 Error probability function showing standard error and probable error.

random will be less than q is equal to the probability that it will be greater than q. That is, there is a 50% probability that the error will lie in the range from $-q$ to $+q$. Stated mathematically,

$$\int_{-q}^{+q} P(r)dr = \frac{1}{2} \tag{12-8}$$

This integral gives the following expression for the probable error of a single measurement:

$$q = 0.675\sqrt{\frac{1}{N-1}\sum r_i^2} = 0.675\sigma \tag{12-9}$$

Let us evaluate the standard error and probable error for the data given in Table 12-1. The mean value of the 50 measurements is 8.84. Using this value, we find the 50 residuals and, from them, determine $\sum r_i^2 = 0.0072$. This gives

$$\sigma = \sqrt{\frac{1}{N-1}\sum r_i^2} = \sqrt{\frac{1}{49}(0.0072)} = \pm 0.012$$

and

$$q = 0.675\sigma = \pm 0.008$$

(Errors generally are expressed to one significant figure.)

Another of the various parameters commonly used by chemists to measure the reliability of a measurement is called the *average error*:

$$\bar{r} = \frac{1}{N}\sum |r_i| \tag{12-10}$$

The advantage of the average error is that it is much easier to calculate than is the standard deviation or the probable error.

We said that when errors are mainly random, the mean value is the best value with a higher precision than that of any single measurement. We find that the error Q associated with the mean value is inversely proportional to the square root of the number of measurements taken:

$$Q = q/\sqrt{N} \tag{12-11}$$

Thus, the probable error of the mean is

$$Q = 0.675\sqrt{\frac{1}{N(N-1)}\sum r_i^2} \tag{12-12}$$

Applying this equation to the set of data given in Table 12-1, we find that

$$Q = \frac{\pm 0.008}{\sqrt{50}} = \pm 0.001$$

Hence, we see that, while any single measurement in the series is reliable to ± 0.008 g, the mean value is reliable to ± 0.001 g:

$$\overline{m} = 8.840 \pm 0.001 \text{ g}$$

Example

Given the following set of measured values, find the mean value, the standard deviation, the probable error of a single measurement, and the probable error of the mean value:

Measurement	*Frequency*
2.334	2
2.335	5
2.336	14
2.337	34
2.338	21
2.339	11
2.340	3
2.341	1

Solution. (There is a macro in Appendix VI to do this calculation with Excel, but the student should try a few of these calculations without using a spreadsheet, in order to see how they work.)

The mean value is 2.337286. (We will let the probable error in the mean determine to how many places past the decimal point we should express the number.) We have

$$\sum r_i^2 = 1.526 \times 10^{-4}$$

$$\sigma = \sqrt{\frac{1}{N-1}\sum r_i^2} = \sqrt{\frac{1}{90}(1.526 \times 10^{-4})} = \pm 0.001302$$

$$q = 0.675\sigma = \pm 8.789 \times 10^{-4} = \pm 0.0009$$

$$Q = \frac{q}{\sqrt{N}} = \frac{0.0009}{\sqrt{91}} = 9.4 \times 10^{-5} = \pm 0.0001$$

$$\text{Therefore, } \overline{m} = 2.3373 \pm 0.0001$$

12-4 PROPAGATION OF ERRORS

Experiments in physical chemistry rarely involve a direct measurement of the physical property in question. For example, in order to determine the density of an object, we normally measure the mass and the volume of the object and relate those quantities to the density. In fact, in many cases, we do not even measure the volume of the object directly. Thus, in experiments such as these, the errors in the measurements are associated with the mass and the volume, and not with the density. Yet, it would be useful to known the reliability of the density, given specific errors in the mass and the volume.

Suppose that some physical property P is a function of several measurable quantities $x_1, x_2, x_3, x_4, \ldots$; that is, let

$$P = f(x_1, x_2, x_3, x_4, \ldots) \tag{12-13}$$

Suppose further that each of these quantities, measured once or a number of times to determine a mean value, has an error Q associated with it. Let Q_1 be the probable error associated with x_1, Q_2 the probable error associated with x_2, and so on. We saw in Chapter 4 that any small change in the variables $x_1, x_2, x_3, \ldots$ will cause a change in P given by the equation

$$dP = \left(\frac{\partial P}{\partial x_1}\right)_{x_2, x_3, \ldots} dx_1 + \left(\frac{\partial P}{\partial x_2}\right)_{x_1, x_3, \ldots} dx_2 + \cdots \tag{12-14}$$

Let us assume that the small changes in the variables are the probable errors in those variables. Hence, the probable error in P can be expressed as

$$Q_P = \left|\frac{\partial P}{\partial x_1}\right| Q_1 + \left|\frac{\partial P}{\partial x_2}\right| Q_2 + \cdots \tag{12-15}$$

where the partial derivatives are evaluated at the mean values if each x variable is measured more than once.

We find, however, that Equation (12-15) does not take into account the fact that random errors tend to cancel each other out. If we square both sides of Equation (12-15), we obtain the equation

$$\begin{aligned} Q_P^2 = {} & \left(\frac{\partial P}{\partial x_1}\right)^2 Q_1^2 + \left(\frac{\partial P}{\partial x_2}\right)^2 Q_2^2 + \cdots \\ & + 2\left(\frac{\partial P}{\partial x_1}\right)\left(\frac{\partial P}{\partial x_2}\right) Q_1 Q_2 + \cdots \end{aligned} \tag{12-16}$$

Because probable errors tend to cancel each other out, the cross terms in Equation (12-16) vanish, giving the equation

$$Q_P^2 = \left(\frac{\partial P}{\partial x_1}\right)^2 Q_1^2 + \left(\frac{\partial P}{\partial x_2}\right)^2 Q_2^2 + \cdots$$

or

$$Q_P = \sqrt{\left(\frac{\partial P}{\partial x_1}\right)^2 Q_1^2 + \left(\frac{\partial P}{\partial x_2}\right)^2 Q_2^2 + \cdots} \tag{12-17}$$

where each partial derivative is evaluated at the mean if each x variable is measured more than once. Equation (12-17) allows us to propagate errors through equations used to calculate the final result.

To illustrate the use of Equation (12-17), consider the following example: Suppose we use a stoppered weighing bottle and an analytical balance to measure the mass of a liquid delivered by a 10-ml pipette. Let us assume that the probable error in any mass measured with this balance is ± 0.0005 g. Assume that the mass of the weighing bottle is 19.9314 ± 0.0005 g and that the mass of the weighing bottle plus the liquid is 28.1038 ± 0.0005 g. The mass of the liquid, therefore, is

$$m_l = m_{l+b} - m_b = 28.1038 - 19.9314 = 8.1724 \text{ g}$$

Since the mass of the liquid was not measured directly, we must determine its probable error with the use of Equation (12-17),

$$Q_{m_l} = \sqrt{\left(\frac{\partial m_l}{\partial m_{l+b}}\right)^2 Q_{m_{l+b}}^2 + \left(\frac{\partial m_l}{\partial m_b}\right)^2 Q_{m_b}^2}$$

Evaluating the partial derivatives yields

$$\left(\frac{\partial m_l}{\partial m_{l+b}}\right) = 1 \quad \text{and} \quad \left(\frac{\partial m_l}{\partial m_b}\right) = -1$$

Thus, we have

$$Q_{m_l} = \sqrt{(1)^2(0.0005)^2 + (-1)^2(0.0005)^2} = \pm 0.0007$$

and

$$m_l = 8.1724 \pm 0.0007 \text{ g}$$

Notice that the error in the mass of the liquid is larger than the error in the mass of the liquid plus bottle and also larger than the error in the mass of the bottle. This is reasonable, since the mass of the liquid was determined by performing two measurements on the balance. Notice further, though, that the error in the mass of the liquid is less than the sum of the error in the mass of the liquid plus bottle and the error in the mass of the bottle. This clearly shows that errors are not purely additive and tend to cancel each other out.

We now can determine the density of the liquid by dividing the mass of the liquid by the volume of the liquid. Let us assume that the error in the 10-ml pipette is ±0.02 ml. Then the density of the liquid is

$$D = \frac{m}{V} = \frac{8.1724}{10.00} = 0.81724 \text{ g/ml}$$

But what is the *error* in the density? To how many significant figures are we justified in expressing the density? We again must use Equation (12-17), this time in the form

$$Q_D = \sqrt{\left(\frac{\partial D}{\partial m}\right)^2 Q_m^2 + \left(\frac{\partial D}{\partial V}\right)^2 Q_V^2}$$

Evaluating the partial derivatives yields

$$\left(\frac{\partial D}{\partial m}\right) = \frac{1}{V} \quad \text{and} \quad \left(\frac{\partial D}{\partial V}\right) = \frac{-m}{V^2}$$

Thus, we can write

$$Q_D = \sqrt{\left(\frac{1}{10.00}\right)^2 (0.0007)^2 + \left(\frac{-8.1724}{(10.00)^2}\right)^2 (0.02)^2} = \pm 0.002$$

Therefore, the density of the liquid can be expressed only to three significant figures as

$$D = 0.817 \pm 0.002 \text{ g/ml}$$

Examples

(a) The density of a liquid is determined by the pycnometric method, using a constant temperature bath at 25.00°C A *pycnometer* is a bottle fitted with a capillary top whose volume is calibrated with a liquid of known density, usually water. The mass error associated with the balance used is ±0.0003 g. Given the following data, find the probable error in the density of the liquid:

Mass of dry pycnometer in air: 23.4621 g
Mass of pycnometer and water in air: 48.7731 g
Mass of pycnometer and liquid in air: 45.8800 g

Take the density of water to be 0.997044 g/cc at 25.00°C, and assume that the error in the density of water is negligible. Also, neglect the effect of air buoyancy for this example.

Solution. We first determine the error in the mass of the liquid. We have

$$m_l = m_{P+l} - m_p = 45.8800 - 23.4621 = 22.4179 \text{ g}$$

$$Q_{m_l} = \sqrt{\left(\frac{\partial m_l}{\partial m_{p+l}}\right)^2 Q_{m_{p+l}}^2 + \left(\frac{\partial m_l}{\partial m_p}\right)^2 Q_{m_p}^2}$$

$$= \sqrt{(1)^2(0.0003)^2 + (-1)^2(0.0003)^2} = \pm 0.0004 \text{ g}$$

Next, we calculate the volume of the liquid and the error in the volume. The volume of the liquid is the volume (capacity) of the pycnometer:

$$V = \frac{m_{p+w} - m_p}{\rho_w} = \frac{48.7731 - 23.4621}{0.997044} = 25.38604 \text{ cc}$$

$$Q_V = \sqrt{\left(\frac{\partial V}{\partial m_{p+w}}\right)^2 Q_{m_{p+w}}^2 + \left(\frac{\partial V}{\partial m_p}\right)^2 Q_{m_p}^2}$$

$$= \sqrt{\left(\frac{1}{\rho_w}\right)^2 (0.0003)^2 + \left(-\frac{1}{\rho_w}\right)^2 (0.0003)^2}$$

$$= \pm 0.0004 \text{ cc. Therefore, } V = 25.3860 \pm 0.0004 \text{ cc}$$

Finally, we calculate the density of the liquid and its error:

$$D = \frac{m_l}{V} = \frac{22.4179}{25.3860} = 0.88308 \text{ g/cc}$$

$$Q_D = \sqrt{\left(\frac{\partial D}{\partial m_l}\right)^2 Q_{m_l}^2 + \left(\frac{\partial D}{\partial V}\right)^2 Q_V^2} = \sqrt{\left(\frac{1}{V}\right)^2 Q_{m_l}^2 + \left(-\frac{m_l}{V^2}\right)^2 Q_V^2}$$

$$= \sqrt{\left(\frac{1}{25.3860}\right)^2 (0.0004)^2 + \left(-\frac{22.4179}{(25.3860)^2}\right)^2 (0.0004)^2}$$

$$= \pm 0.00002 \text{ g/cc}$$

Therefore, the density of the liquid is 0.88308 ± 0.00002 g/cc.

(b) A rectangular box has the following measured sides: $a = 10.123 \pm 0.003$ cm, $b = 4.677 \pm 0.003$ cm, and $c = 6.987 \pm 0.003$ cm. Find the probable error in the volume of the box.

Solution. We have

$$V = abc; \; Q_V = \sqrt{\left(\frac{\partial V}{\partial a}\right)^2 Q_a^2 + \left(\frac{\partial V}{\partial b}\right)^2 Q_b^2 + \left(\frac{\partial V}{\partial c}\right)^2 Q_c^2}$$

$$V = (10.123)(4.677)(6.987) = 330.8014 \text{ cc}$$

$$Q_V = \sqrt{(bc)^2(0.003)^2 + (ac)^2(0.003)^2 + (ab)^2(0.003)^2}$$

$$= \sqrt{(32.6781)^2(0.003)^2 + (70.7294)^2(0.003)^2 + (47.3452)^2(0.003)^2}$$

$$= \pm 0.3 \text{ cc}$$

Therefore, the volume of the box is $V = 330.8 \pm 0.3$ cc.

12-5 PREPARATION OF GRAPHS

We saw in Chapter 2 that one of the most useful ways to display the dependence of one function on another is by means of a graph. In Chapter 11, we discussed the use of Excel for the preparation of graphs, and students are encouraged to use the Chart Wizard feature of the spreadsheet to prepare graphs. Spreadsheets, however, are designed for the scientist as well as the nonscientist; therefore, science students should know something about how their graphs are supposed to look.

First, choose a suitable set of coordinate axes. Be sure to mark and number the main divisions along each axis clearly and to choose a correct scale so that important data do not run off the graph. Label each axis, and include the units of the measured values in parentheses after the label—for example, *P*(bar), *V*(cc), or *T*(K). It is customary also to include a title that describes the data being plotted (e.g., *Pressure versus Volume for* CO_2 *Gas*) or the equation of the curve being plotted (for example, $\ln k = -E_a/RT + \ln A$). Excel offers a wide choice for the designation of the points, although circles are the most common. Remember to use the *XY*-scatter type of graph.

Experimental data normally are continuous; hence, data points should be connected with a *smooth* curve through them. Avoid the feature of Chart Wizard that connects the points together with a series of short, straight lines ("follow the dots"). If the equation of the curve is known, or if the data points are to be fitted to some specific type of equation, such as a linear equation, with *Trendline*, then plot the graph using "Scatter—compares pairs of points," and have *Trendline* plot the graph.

For some experiments in physical chemistry, you may not find it convenient to use Excel. If you plot a graph "from scratch," use the following method: Choose a good grade of graph paper. (Notebook paper is *not* acceptable!) Engineering graph paper divided 10 × 10 to the centimeter is suitable for most physical chemistry experiments. Place the data points on the graph paper with a sharp pencil or a common pin. Sketch the curve lightly with a hard-lead pencil. Do *not* sketch the curves freehand. If the set of points is supposed to be linear, use a straightedge to draw the line. If the set of points is supposed to fall on a curve, use a French curve—a plastic

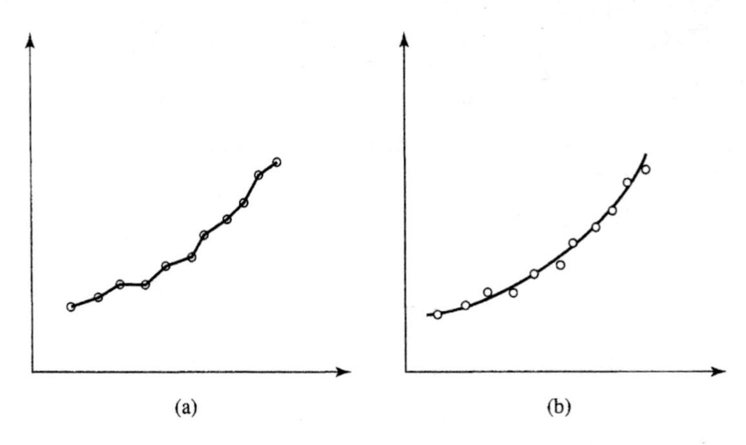

Figure 12-4 Graphical representation of a typical set of data points: (a) incorrect method of connecting points; (b) correct method of drawing smooth curve through a set of points.

device containing several irregular curves that can act as a template—to draw the curve. Again, do not play "follow the dots." It is not necessary that the curve pass through all the points. It is more important that the curve follow the trend of the points smoothly, as illustrated in Fig. 12-4. Do not be afraid to erase what you have drawn if you are not satisfied with the curve. Once you have drawn the best curve through the points, you may trace over the curve with ink (again, using a straight-edge or a French curve) and erase the penciled lines with a soft eraser. Some laboratory texts suggest not passing the curves through the circles, but only up to their outer edge, since the curve will obscure the experimental points.

12-6 LINEAR REGRESSION

One of the best methods for obtaining physical constants from experimental curves is by using the *method of least squares*. Although this method can be used for curves of higher order, we shall consider it only as applied to first-degree, or linear, equations. We saw in an earlier chapter that we can determine the slope and the y-intercept of a line using the *Trendline* feature of Excel. While this practice is extremely useful and should be employed whenever possible, it also is important for students of physical chemistry to understand the mathematics behind linear regression.

Recall from Chapter 2 that a linear equation has the form

$$y = mx + b$$

where m is the slope of the line and b is the y-intercept. Let us assume that, for every point (x_i, y_i) in a series of points, due to experimental error there is a deviation from linearity of

$$s_i = y_i - m_i - b$$

The square of this deviation is

$$\sigma_i = s_i^2 = y_i^2 - 2mx_iy_i - 2y_ib + m^2x_i^2 + 2mx_ib + b^2$$

and the total squared deviation is the sum of the squared deviations:

$$\sigma = \sum\sigma_i = \sum y_i^2 - \sum 2mx_iy_i - \sum 2y_ib + \sum m^2x_i^2 + \sum 2mx_ib + \sum b^2 \tag{12-18}$$

The method of least squares requires that the best straight line drawn through a series of points be a straight line in which σ is a minimum with respect to m and b; that is,

$$\frac{\partial\sigma}{\partial m} = \frac{\partial\sigma}{\partial b} = 0$$

Taking the partial derivatives, we have

$$\frac{\partial\sigma}{\partial m} = -\sum 2x_iy_i + \sum 2mx_i^2 + \sum 2x_ib = 0 \tag{12-19}$$

$$\frac{\partial\sigma}{\partial b} = -\sum 2y_i + \sum 2mx_i + \sum 2b = 0 \tag{12-20}$$

Equations (12-19) and (12-20) can be solved simultaneously for m and b by Cramer's rule. To set up the equations for Cramer's rule, we write them as

$$m\sum 2x_i^2 + b\sum 2x_i = \sum 2x_iy_i$$

$$m\sum 2x_i + b\sum 2 = \sum 2y_i$$

(Note that $\sum 2$ means that we add 2 for each data point.) Using Cramer's rule, we see that

$$D = \begin{vmatrix} \sum 2x_i^2 & \sum 2x_i \\ \sum 2x_i & \sum 2 \end{vmatrix}, \quad D_1 = \begin{vmatrix} \sum 2x_iy_i & \sum 2x_i \\ \sum 2y_i & \sum 2 \end{vmatrix},$$

$$D_2 = \begin{vmatrix} \sum 2x_i^2 & \sum 2x_iy_i \\ \sum 2x_i & \sum 2y_i \end{vmatrix},$$

TABLE 12-2 DATA POINTS FOR LINEAR REGRESSION

x_i	y_i	$2x_i$	$2x_i^2$	$2x_iy_i$	$2y_i$
1.7911	23.226	3.5822	6.4161	83.2001	46.452
1.2500	21.195	2.5000	3.1250	52.9875	42.390
0.8736	19.609	1.7472	1.5264	34.2608	39.218
0.6177	18.793	1.2354	0.7631	23.2169	37.586
0.4368	18.096	0.8736	0.3816	15.8087	36.192
	Sums:	9.9384	12.2122	209.4740	201.838

$$m = \frac{D_1}{D}, \quad \text{and} \quad b = \frac{D_2}{D}$$

Example

Given the following set of linear data points, find the best slope and y-intercept, using the method of least squares:

To solve the problem, we set up a table such as Table 12-2.

$$D = \begin{vmatrix} 12.2122 & 9.9384 \\ 9.9384 & 10 \end{vmatrix} = 23.350 \quad D_1 = \begin{vmatrix} 209.474 & 9.9384 \\ 201.838 & 10 \end{vmatrix} = 88.793$$

$$D_2 = \begin{vmatrix} 12.2122 & 209.474 \\ 9.9384 & 201.838 \end{vmatrix} = 383.050$$

$$m = \frac{88.793}{23.350} = 3.8026 \quad \text{and} \quad b = \frac{383.050}{23.350} = 16.404$$

12-7 TANGENTS AND AREAS

Many experiments in physical chemistry require us to extract information from the slopes of the curves we plot. Unfortunately, it is not always possible to plot data in such a way that straight lines will result (which, of course, would make the determination of the slope very easy). Recall, however, that the slope of a curve at a point is the slope of a line drawn tangent to the curve at that point. If the equation of the curve is known, we can determine the slope of the line from its first derivative. If the equation of the curve is not known, the curve still may be able to be fit to an equation using the *Trendline* feature of Excel.

One method of constructing tangents to curves is known as the *method of chords*. With a compass placed at the point in question on the curve, strike off two

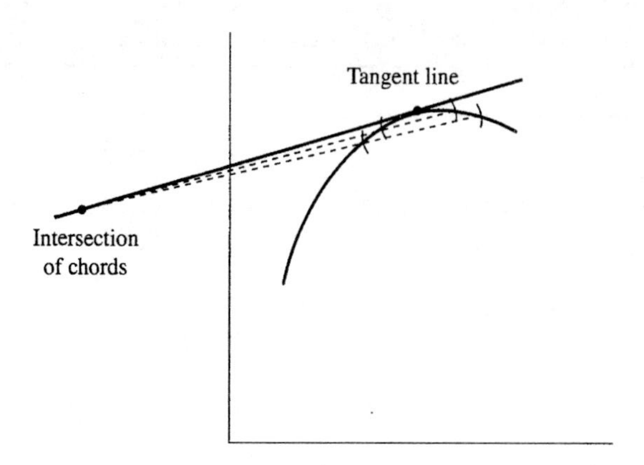

Figure 12-5 Construction of tangent line by method of chords.

arcs, one on either side of the point, as illustrated in Fig. 12-5. Next, draw chords through the intersection of the arcs with the curve until the chords intersect. A line drawn from the intersection of the two chords to the point on the curve is a good approximation of the tangent to the curve at that point.

It also may be necessary in some experiments to determine the areas under curves graphically. This is generally the case in spectroscopy, where the intensity of a spectral line is equal to the area under the curve. Such areas are commonly graphed in gas chromatography as well, where the area under the chromatograph peak is directly proportional to the amount of substance causing the peak. There are several acceptable ways to determine an area under a curve. One method, known as the Riemann sum approximation, is to divide the area into a series of rectangles, as shown in Fig. 12-6, and then measure the area of each rectangle. The rectangles are chosen so that the small triangular area above the curve is approximately equal to the small triangular area below the curve.

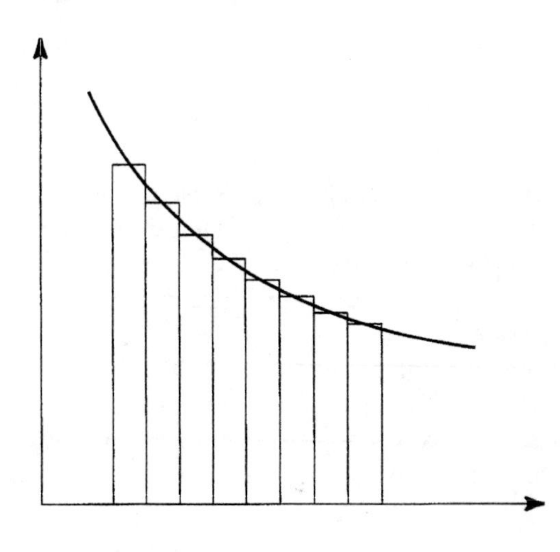

Figure 12-6 Determination of area under curve by the Riemann sum approximation. (Rectangles are chosen so that the triangular areas above the curve approximately equal the triangular areas below the curve. Obviously, the smaller the rectangles, the closer the match between area of the rectangles and the actual area under the curve.)

Another method, which is quite accurate if done correctly, is to cut out the area and determine its mass on an analytical balance. Then, the mass of the area cut out is compared with the mass of a known area on the same graph paper. For this method to work, it is important that a good grade of graph paper having a uniform thickness be used.

A quick method for determining the area under the curve graphically is to use a planimeter, an instrument that is mechanically run around the boundaries of the area. Although this method is fast, it is not accurate if the area to be determined is small.

Many commercial instruments that display data graphically, such as magnetic resonance spectrometers, have built-in electronic integrators that will automatically give the areas under curves. As more and more commercial instrumentation is interfaced with computers, the task of analyzing the data produced by the instrumentation will become easier.

PROBLEMS

1. Determine the probability of throwing a 2, 3, 4, 5, 6, 7, and 11 with a pair of honest dice.

2. Suppose you are given the following set of data:

Measurement	*Frequency*
5.61	2
5.62	11
5.63	18
5.64	30
5.65	35
5.66	27
5.67	14
5.68	9
5.69	4

Determine

(a) the arithmetic mean

(b) the standard deviation

(c) the probable error of a single value

(d) the probable error of the mean

(e) the average error

3. Repeat the calculations asked for in Problem 2, using the macro Statistics() found in Appendix VI.

4. The volume of a cylindrical capillary tube is given by the expression $V = \pi r^2 h$, where r is the radius of the tube and h is the length of the tube. If the radius of the capillary tube is found to be 0.030 cm with a probable error of ± 0.002 cm and the length of the tube is found to be 4.0 cm with a probable error of ± 0.1 cm, what is the volume of the capillary tube and its probable error? What measurement must be made to a higher precision to decrease the probable error in the volume?

5. The molar mass of a vapor is determined by filling a bulb of known volume with the vapor at a known temperature and pressure and measuring the mass of the bulb. This method is known as the *Dumas method.* If the vapor is assumed to be an ideal gas, then, from the ideal gas law,

$$M = \frac{mRT}{PV}$$

where M is the molar mass, m is the mass of the vapor, R is the gas constant, T is the absolute temperature, P is the pressure, and V is the volume. Given that

$$m = 1.0339 \pm 0.0007 \text{ g}$$
$$T = 274.0 \pm 0.5 \text{ K}$$
$$P = 1.050 \pm 0.001 \text{ bar}$$
$$V = 0.1993 \pm 0.0001 \text{ liters}$$
$$R = 0.08314\ \ell \cdot \text{bar} \cdot \text{mol}^{-1} \cdot \text{K}^{-1} \text{ (no probable error)}$$

determine the error in the molar mass.

6. On millimeter graph paper (using an expanded scale), plot the curve $y = x^3 - 3x + x + 1$ from $x = -1$ to $x = +1$. Using the method of chords, find the slope of the curve at $x = 0$. Compare the slope found by this method with that found by using differential calculus.
7. On uniform graph paper (using an expanded scale), plot the curve $y = \frac{1}{2}x^2$ from $x = 0$ to $x = 4$. Determine the area under the curve from $x = 1$ to $x = 3$ by cutting out the area and determining its mass on an analytical balance. Next, determine the area by breaking up the area into small rectangles and determining the total area of the rectangles. Compare the areas found by these methods with the actual area found by integral calculus.
8. Prepare a graph of the data given in Problem 11-6, and determine the change in enthalpy by graphical integration.
9. Prepare a graph of the data given in Problem 11-7, and determine the change in entropy by graphical integration.

Appendices

I TABLE OF PHYSICAL CONSTANTS

Constant	Symbol	Value (SI Units)
Avogadro's number	N_0	$6.022169 \times 10^{23}\ \text{mol}^{-1}$
Boltzmann constant	k	$1.380662 \times 10^{-23}\ \text{J} \cdot \text{molecule}^{-1} \cdot \text{K}^{-1}$
Electron rest mass	m_e	$9.109558 \times 10^{-31}\ \text{kg}$
Electron charge	e	$1.602191 \times 10^{-19}\ \text{C}$
Faraday constant	F	$9.648670 \times 10^{4}\ \text{C} \cdot \text{mol}^{-1}$
Gas constant	R	$8.31434 \times 10^{0}\ \text{J} \cdot \text{mol}^{-1} \cdot \text{K}^{-1}$
Permittivity of vacuum	ε_0	$8.854188 \times 10^{-12}\ \text{C}^2 \cdot \text{J}^{-1} \cdot \text{m}$
Planck's constant	h	$6.626196 \times 10^{-34}\ \text{J} \cdot \text{s}$
Proton rest mass	m_p	$1.672614 \times 10^{-27}\ \text{kg}$
Rydberg constant	R_H	$1.096776 \times 10^{7}\ \text{m}^{-1}$
Speed of light in a vacuum	c	$2.997925 \times 10^{8}\ \text{m} \cdot \text{s}^{-1}$

II TABLE OF INTEGRALS

A. Indefinite Integrals[1]

Algebraic Functions

1. $$\int x^p\,dx = \frac{1}{p+1}x^{p+1} + C, \quad \text{if } p \neq -1$$

2. $$\int \frac{dx}{x} = \ln|x| + C$$

Algebraic Functions of x and ax + b

3. $$\int (ax+b)^p\,dx = \frac{(ax+b)^{p+1}}{a(p+1)} + C, \quad \text{if } p \neq -1$$

4. $$\int \frac{dx}{ax+b} = \frac{1}{a}\ln|ax+b| + C$$

[1]From E. J. Cogan and R. Z. Norman, *Handbook of Calculus, Difference, and Differential Equations*, Englewood Cliffs, NJ: Prentice-Hall, 1958, pp. 185–201.

5. $$\int \frac{x\,dx}{ax+b} = \frac{1}{a^2}[ax - b\,\mathbf{ln}|ax+b|] + C$$

6. $$\int \frac{x\,dx}{(ax+b)^2} = \frac{1}{a^2}\left(\mathbf{ln}|ax+b| + \frac{b}{ax+b}\right) + C$$

7. $$\int \frac{x\,dx}{(ax+b)^3} = \frac{1}{a^2}\left(\frac{b}{2(ax+b)^2} - \frac{1}{ax+b}\right) + C$$

8. $$\int x(ax+b)^p\,dx = \frac{1}{a^2(p+2)}(ax+b)^{p+2} - \frac{b}{a^2(p+1)}(ax+b)^{p+1} + C, \quad \text{if } p \neq -1, p \neq -2$$

9. $$\int x^m(ax+b)^n\,dx = \frac{1}{m+n+1}[x^{m+1}(ax+b)^n + nb\int x^m(ax+b)^{n-1}\,dx], \quad \text{if } m+n+1 \neq 0$$

See (1).

10. $$\int x^m(ax+b)^n\,dx = \frac{1}{a(m+n+1)}[x^m(ax+b)^{n+1} - mb\int x^{m-1}(ax+b)^n\,dx], \quad \text{if } m+n+1 \neq 0$$

See (3).

11. $$\int \frac{x^2\,dx}{ax+b} = \frac{1}{a^3}\left(\frac{1}{2}(ax+b)^2 - 2b(ax+b) + b^2\,\mathbf{ln}|ax+b|\right) + C$$

12. $$\int \frac{x^2\,dx}{(ax+b)^2} = \frac{1}{a^3}\left(ax+b - 2b\,\mathbf{ln}|ax+b| - \frac{b^2}{ax+b}\right) + C$$

13. $$\int \sqrt{ax+b}\,dx = \frac{2}{3a}\sqrt{(ax+b)^3} + C$$

14. $$\int \frac{dx}{\sqrt{ax+b}} = \frac{2\sqrt{ax+b}}{a} + C$$

15. $$\int x\sqrt{ax+b}\,dx = \frac{2(3ax-2b)\sqrt{(ax+b)^3}}{15a^2} + C$$

16. $$\int \frac{\sqrt{ax+b}}{x}\,dx = 2\sqrt{ax+b} + b\int \frac{dx}{x\sqrt{ax+b}}$$

See (17) if $b > 0$; (18) if $b < 0$.

17. $$\int \frac{dx}{x\sqrt{ax+b}} = \frac{1}{\sqrt{b}}\,\mathbf{ln}\left|\frac{\sqrt{ax+b}-\sqrt{b}}{\sqrt{ax+b}+\sqrt{b}}\right| + C = -\frac{1}{\sqrt{b}}\,\mathbf{tanh}^{-1}\sqrt{\frac{ax+b}{b}} + C, \quad \text{if } b > 0$$

18. $$\int \frac{dx}{x\sqrt{ax-b}} = \frac{2}{\sqrt{b}}\,\mathbf{arctan}\sqrt{\frac{ax-b}{b}} + C, \quad \text{if } b > 0$$

19. $\int x^2\sqrt{ax+b}\,dx = \frac{2}{105a^3}\sqrt{(ax+b)^3}(15a^2x^2 - 12abx + 8b^2) + C$

20. $\int \frac{x^n\,dx}{\sqrt{ax+b}} = \frac{2}{a(2n+1)}\left[x^n\sqrt{ax+b} - nb\int \frac{x^{n-1}\,dx}{\sqrt{ax+b}}\right]$

21. $\int \frac{dx}{x^n\sqrt{ax+b}} = -\frac{\sqrt{ax+b}}{(n-1)bx^{n-1}} - \frac{(2n-3)a}{(2n-2)b}\int \frac{dx}{x^{n-1}\sqrt{ax+b}}$

See (14).

Algebraic Functions of x and $ax^2 + b$

22. $\int \frac{dx}{ax^2+b} = \frac{1}{\sqrt{ab}}\,\mathbf{arctan}\frac{x\sqrt{ab}}{b} + C, \quad \text{if } ab > 0$

23. $\int \frac{dx}{ax^2+b} = \frac{1}{2\sqrt{-ab}}\,\mathbf{ln}\left|\frac{b + x\sqrt{-ab}}{b - x\sqrt{-ab}}\right| + C$

$= \frac{1}{\sqrt{-ab}}\,\mathbf{tanh}^{-1}\frac{x\sqrt{-ab}}{b} + C, \quad \text{if } ab < 0$

24. $\int \frac{dx}{ax^2-b} = \frac{1}{2\sqrt{ab}}\,\mathbf{ln}\left|\frac{x\sqrt{a} - \sqrt{b}}{x\sqrt{a} + \sqrt{b}}\right| + C, \quad \text{if } a > 0, b > 0$

25. $\int \frac{dx}{-ax^2+b} = \frac{1}{2\sqrt{ab}}\,\mathbf{ln}\left|\frac{\sqrt{b} + x\sqrt{a}}{\sqrt{b} - x\sqrt{a}}\right| + C, \quad \text{if } a > 0, b > 0$

26. $\int \frac{dx}{(ax^2+b)^{m+1}} = \frac{1}{2mb}\left[\frac{x}{(ax^2+b)^m} + (2m-1)\int \frac{dx}{(ax^2+b)^m}\right]$

See (23) or (24) if $b > 0$; (23) if $b < 0$.

27. $\int x(ax^2+b)^p\,dx = \frac{1}{2a}\frac{(ax^2+b)^{p+1}}{p+1} + C, \quad \text{if } p \neq -1$

28. $\int \frac{x\,dx}{ax^2+b} = \frac{1}{2a}\,\mathbf{ln}|ax^2+b| + C$

29. $\int \frac{x\,dx}{(ax^2+b)^{p+1}} = \frac{-1}{2pa(ax^2+b)^p} + C, \quad \text{if } p > 0$

30. $\int \frac{dx}{x(ax^n+b)} = \frac{1}{bn}\,\mathbf{ln}\left|\frac{x^n}{ax^n+b}\right| + C$

31. $\int x^p(ax^n+b)^m\,dx = \frac{1}{nm+p+1}[x^{p+1}(ax^n+b)^m$

$+ bnm\int x^p(ax^n+b)^{m-1}\,dx],$

if $m > 0$ and $mn + p + 1 \neq 0$

See (1).

32. $\int \sqrt{ax^2+b}\,dx = \frac{x}{2}\sqrt{ax^2+b} + \frac{b}{2\sqrt{a}}\,\mathbf{ln}|x\sqrt{a}$

$+ \sqrt{ax^2+b}| + C, \quad \text{if } a > 0$

33. $$\int \sqrt{ax^2 + b}\, dx = \frac{x}{2}\sqrt{ax^2 + b} + \frac{b}{2\sqrt{-a}} \mathbf{arcsin}\left(x\sqrt{\frac{-a}{b}}\right) + C, \quad \text{if } a < 0$$

Algebraic Functions of x and $ax^2 + bx + c$

$$X = ax^2 + bx + c, q = 4ac - b^2$$

34. $$\int \frac{dx}{ax^2 + bx + c} = \frac{2}{\sqrt{q}} \mathbf{arctan} \frac{2ax + b}{\sqrt{q}} + C, \quad q > 0$$

35. $$\int \frac{dx}{ax^2 + bx + c} = \frac{2}{\sqrt{-q}} \mathbf{ln} \left| \frac{2ax + b - \sqrt{-q}}{2ax + b + \sqrt{-q}} \right| + C$$

$$= \frac{-2}{\sqrt{-q}} \mathbf{tanh}^{-1} \frac{2ax + b}{\sqrt{-q}} + C, \quad q < 0$$

36. $$\int \frac{dx}{(ax^2 + bx + c)^{n+1}} = \frac{2ax + b}{nqX^n} + \frac{2(2n - 1)a}{nq} \int \frac{dx}{X^n}$$

See (34) if $q > 0$; (35) if $q < 0$.

37. $$\int \sqrt{ax^2 + bx + c}\, dx = \frac{2ax + b}{4a}\sqrt{X} + \frac{q}{8a} \int \frac{dx}{\sqrt{X}}$$

See (38) if $a > 0$; (39) if $a < 0$.

38. $$\int \frac{dx}{\sqrt{ax^2 + bx + c}} = \frac{1}{\sqrt{a}} \mathbf{ln} \left| X + x\sqrt{a} + \frac{b}{2\sqrt{a}} \right| + C, \quad \text{if } a > 0$$

39. $$\int \frac{dx}{\sqrt{ax^2 + bx + c}} = \frac{1}{\sqrt{-a}} \mathbf{arcsin} \frac{-2ax - b}{\sqrt{-q}} + C, \quad \text{if } a < 0$$

40. $$\int \frac{x\, dx}{ax^2 + bx + c} = \frac{1}{2a} \left[\mathbf{ln}|X| - b \int \frac{dx}{X} \right]$$

See (32) if $q > 0$; (33) if $q < 0$.

Algebraic Functions of x and $x^2 \pm a^2$

41. $$\int \frac{dx}{x^2 + a^2} = \frac{1}{a} \mathbf{arctan} \frac{x}{a} + C$$

42. $$\int \frac{dx}{x^2 - a^2} = \frac{1}{2a} \mathbf{ln} \left| \frac{x - a}{x + a} \right| + C = -\frac{1}{a} \mathbf{coth}^{-1} \frac{x}{a} + C$$

43. $$\int \sqrt{x^2 \pm a^2}\, dx = \frac{1}{2}\left[x\sqrt{x^2 \pm a^2} \pm a^2 \mathbf{ln} \left| x + \sqrt{x^2 \pm a^2} \right| \right] + C$$

44. $$\int \frac{dx}{\sqrt{x^2 \pm a^2}} = \mathbf{ln}\left|x + \sqrt{x^2 \pm a^2}\right| + C$$

45. $$\int x\sqrt{x^2 \pm a^2}\, dx = \frac{1}{3}\sqrt{(x^2 \pm a^2)^3} + C$$

46. $$\int \frac{\sqrt{x^2 + a^2}}{x} dx = \sqrt{x^2 + a^2} - a \ln\left|\frac{a + \sqrt{x^2 + a^2}}{x}\right| + C$$

47. $$\int \frac{\sqrt{x^2 - a^2}}{x} dx = \sqrt{x^2 - a^2} - a \operatorname{arcsec}\frac{x}{a} + C$$

48. $$\int \frac{x\, dx}{\sqrt{x^2 \pm a^2}} = \sqrt{x^2 \pm a^2} + C$$

49. $$\int \frac{dx}{x\sqrt{x^2 + a^2}} = -\frac{1}{a}\ln\left|\frac{\sqrt{x^2 + a^2} + a}{x}\right| + C = -\frac{1}{a}\sinh^{-1}\frac{a}{x} + C$$

50. $$\int \frac{dx}{x\sqrt{x^2 - a^2}} = \frac{1}{a}\operatorname{arcsec}\frac{x}{a} + C = \frac{1}{a}\arccos\frac{a}{x} + C$$

51. $$\int \sqrt{(x^2 \pm a^2)^3}\, dx = \frac{1}{8}\left[2x\sqrt{(x^2 \pm a^2)^3} \pm 3a^2x\sqrt{x^2 \pm a^2} + 3a^4 \ln\left|x - \sqrt{x^2 \pm a^2}\right|\right] + C$$

52. $$\int \frac{dx}{\sqrt{(x^2 \pm a^2)^3}} = \frac{\pm x}{a^2\sqrt{x^2 \pm a^2}} + C$$

53. $$\int x\sqrt{(x^2 \pm a^2)^3}\, dx = \frac{1}{5}\sqrt{(x^2 \pm a^2)^5} + C$$

54. $$\int \frac{x\, dx}{(x^2 \pm a^2)^3} = \frac{-1}{\sqrt{x^2 \pm a^2}} + C$$

Algebraic Functions of x and $a^2 - x^2$

55. $$\int \frac{dx}{a^2 - x^2} = \frac{1}{2a}\ln\left|\frac{a + x}{a - x}\right| + C = \frac{1}{a}\tanh^{-1}\frac{x}{a} + C$$

56. $$\int \sqrt{a^2 - x^2}\, dx = \frac{1}{2}\left[x\sqrt{a^2 - x^2} + a^2 \arcsin\frac{x}{a}\right] + C$$

57. $$\int \frac{dx}{\sqrt{a^2 - x^2}} = \arcsin\frac{x}{a} + C = -\arccos\frac{x}{a} + C$$

58. $$\int x\sqrt{a^2 - x^2}\, dx = -\frac{1}{3}\sqrt{(a^2 - x^2)^3} + C$$

59. $$\int \frac{\sqrt{a^2 - x^2}}{x} dx = \sqrt{a^2 - x^2} - a \ln\left|\frac{a + \sqrt{a^2 - x^2}}{x}\right| + C$$

60. $$\int \frac{x\, dx}{\sqrt{a^2 - x^2}} = -\sqrt{a^2 - x^2} + C$$

61. $$\int \frac{dx}{x\sqrt{a^2 - x^2}} = -\frac{1}{a}\ln\left|\frac{a + \sqrt{a^2 - x^2}}{x}\right| + C = -\frac{1}{a}\cosh^{-1}\frac{a}{x} + C$$

62. $$\int \sqrt{(a^2 - x^2)^3}\, dx = \frac{1}{8}\left[2x\sqrt{(a^2 - x^2)^3} + 3a^2x\sqrt{a^2 - x^2} + 3a^4 \arcsin\frac{x}{a}\right] + C$$

63. $$\int \frac{dx}{\sqrt{(a^2 - x^2)^3}} = \frac{x}{a^2\sqrt{a^2 - x^2}} + C$$

64. $$\int x\sqrt{(a^2 - x^2)^3}\, dx = -\frac{1}{5}\sqrt{(a^2 - x^2)^5} + C$$

65. $$\int \frac{x\, dx}{\sqrt{(a^2 - x^2)^3}} = \frac{1}{\sqrt{a^2 - x^2}} + C$$

Other Algebraic Functions

66. $$\int \sqrt{2ax - x^2}\, dx = \frac{1}{2}\left[(x - a)\sqrt{2ax - x^2} + a^2 \arcsin\frac{x - a}{a}\right] + C$$

67. $$\int \frac{dx}{\sqrt{2ax - x^2}} = \arccos\frac{a - x}{a} + C$$

68. $$\int \sqrt{\frac{1 + x}{1 - x}}\, dx = \arcsin x - \sqrt{1 - x^2} + C$$

Trigonometric Functions

69. $$\int \sin x\, dx = -\cos x + C$$

70. $$\int \sin(ax + b)\, dx = -\frac{1}{a}\cos(ax + b) + C$$

71. $$\int \sin^2(ax + b)\, dx = \frac{x}{2} - \frac{1}{2a}\cos(ax + b)\sin(ax + b) + C = \frac{x}{2} - \frac{\sin 2(ax + b)}{4a} + C$$

72. $$\int \sin^3(ax + b)\, dx = -\frac{1}{3a}\cos(ax + b)[\sin^2(ax + b) + 2] + C$$

73. $$\int \sin^4(ax + b)\, dx = \frac{3x}{8} - \frac{3\sin 2(ax + b)}{16a} - \frac{\sin^3(ax + b)\cos(ax + b)}{4a} + C$$

74. $$\int \sin^n(ax + b)\, dx = -\frac{1}{an}\left[\sin^{n-1}(ax + b)\cos(ax + b) - a(n - 1)\int \sin^{n-2}(ax + b)\, dx\right]$$

See (73) if n is even; (72) if n is odd.

75. $$\int x \sin(ax+b)\,dx = \frac{1}{a^2}\sin(ax+b) - \frac{x}{a}\cos(ax+b) + C$$

76. $$\int x \sin^2(ax+b)\,dx = \frac{x^2}{4} - \frac{x\sin 2(ax+b)}{4a} - \frac{\cos 2(ax+b)}{8a^2} + C$$

77. $$\int x^2 \sin^2(ax+b)\,dx = \frac{x^3}{6} - \left(\frac{x^2}{4a} - \frac{1}{8a^3}\right)\sin 2(ax+b) - \frac{x\cos 2(ax+b)}{4a^2} + C$$

78. $$\int x^n \sin(ax+b)\,dx = -\frac{1}{a}\left[x^n\cos(ax+b) - n\int x^{n-1}\cos(ax+b)\,dx\right]$$

See (91) if n is even; (75) if n is odd.

79. $$\int \frac{\sin ax\,dx}{x} = ax - \frac{(ax)^3}{3(3!)} + \frac{(ax)^5}{5(5!)} - + \cdots + C$$

80. $$\int \frac{\sin(ax+b)\,dx}{x^n} = \frac{-1}{n-1}\,\frac{\sin(ax+b)}{x^{n-1}} + \frac{a}{n-1}\int\frac{\cos(ax+b)\,dx}{x^{n-1}}, \quad \text{if } n>1$$

See (70) if n is even; (86) if n is odd.

81. $$\int \frac{dx}{1\pm\sin(ax+b)} = \mp\frac{1}{a}\tan\left(\frac{\pi}{4}\mp\frac{ax+b}{2}\right) + C$$

82. $$\int \sqrt{1+\sin x}\,dx = \pm 2\left[\sin\frac{x}{2} - \cos\frac{x}{2}\right] + C$$

Use + sign if $(8n-1)\frac{\pi}{2} < x \le (8n+3)\frac{\pi}{2}$ for some integer n; otherwise use − sign.

83. $$\int \sqrt{1-\sin x}\,dx = \pm 2\left[\sin\frac{x}{2} + \cos\frac{x}{2}\right] + C$$

Use + sign if $(8n-3)\frac{\pi}{2} < x \le (8n+1)\frac{\pi}{2}$ for some integer n; otherwise use − sign.

84. $$\int \sin ax \sin bx\,dx = \frac{\sin(a-b)x}{2(a-b)} - \frac{\sin(a+b)x}{2(a+b)} + C, \quad \text{if } a^2 \ne b^2$$

85. $$\int \cos x\,dx = \sin x + C$$

86. $$\int \cos(ax+b)\,dx = \frac{1}{a}\sin(ax+b) + C$$

87. $$\int \cos^2(ax+b)\,dx = \frac{x}{2} + \frac{1}{2a}\cos(ax+b)\sin(ax+b) + C = \frac{x}{2} + \frac{\sin 2(ax+b)}{4a} + C$$

88. $\int \cos^3(ax+b)\,dx = \frac{1}{a}\sin(ax+b) - \frac{1}{3a}\sin^3(ax+b) + C$

89. $\int \cos^4(ax+b)\,dx = \frac{3x}{8} + \frac{3\sin 2(ax+b)}{16a} + \frac{\cos^3(ax+b)\sin(ax+b)}{4a} + C$

90. $\int \cos^n(ax+b)\,dx = \frac{1}{an}\left[\cos^{n-1}(ax+b)\sin(ax+b)\right.$

$\left. + a(n-1)\int \cos^{n-2}(ax+b)\,dx\right]$

See (88) if n is odd; (89) if n is even.

91. $\int x\cos(ax+b)\,dx = \frac{1}{a^2}\cos(ax+b) + \frac{x}{2}\sin(ax+b) + C$

92. $\int x\cos^2(ax+b)\,dx = \frac{x^2}{4} + \frac{x\sin 2(ax+b)}{4a} + \frac{\cos 2(ax+b)}{8a^2} + C$

93. $\int x^2\cos^2(ax+b)\,dx = \frac{x^3}{6} + \left(\frac{x^2}{4a} - \frac{1}{8a^3}\right)\sin 2(ax+b)$

$+ \frac{x\cos 2(ax+b)}{4a^2} + C$

94. $\int x^n\cos(ax+b)\,dx = \frac{1}{a}\left[x^n\sin(ax+b) - n\int x^{n-1}\sin(ax+b)\,dx\right] + C$

See (75) if n is even; (91) if n is odd.

95. $\int \frac{\cos ax\,dx}{x} = \ln|ax| - \frac{(ax)^2}{2(2!)} + \frac{(ax)^4}{4(4!)} - + \cdots + C$

96. $\int \frac{\cos(ax+b)\,dx}{x^n} = \frac{-1}{n-1}\frac{\cos(ax+b)}{x^{n-1}}$

$- \frac{a}{n-1}\int \frac{\sin(ax+b)\,dx}{x^{n-1}}, \quad \text{if } n > 1$

See (86) if n is even; (70) if n is odd.

97. $\int \frac{dx}{1+\cos(ax+b)} = \frac{1}{a}\tan\left(\frac{ax+b}{2}\right) + C$

98. $\int \frac{dx}{1-\cos(ax+b)} = -\frac{1}{a}\cot\left(\frac{ax+b}{2}\right) + C$

99. $\int \sqrt{1+\cos x}\,dx = \pm 2\sqrt{2}\sin\frac{x}{2} + C$

Use + sign if $(4n-1)\pi < x \leq (4n+1)\pi$ for some integer n; otherwise use − sign.

100. $\int \sqrt{1-\cos x}\,dx = \pm 2\sqrt{2}\cos\frac{x}{2} + C$

Use + sign if $(4n-2)\pi < x \leq 4n\pi$ for some integer n; otherwise use − sign.

101. $\int \cos ax\cos bx\,dx = \frac{\sin(a-b)x}{2(a-b)} + \frac{\sin(a+b)x}{2(a+b)} + C, \quad \text{if } a^2 \neq b^2$

102. $$\int \sin ax \cos bx\, dx = -\frac{\cos(a-b)x}{2(a-b)} - \frac{\cos(a+b)x}{2(a+b)} + C, \quad \text{if } a^2 \neq b^2$$

103. $$\int \sin(ax+b)\cos(ax+b)\, dx = \frac{1}{2a}\sin^2(ax+b) + C$$

104. $$\int \sin^p(ax+b)\cos(ax+b)\, dx = \frac{1}{a(p+1)}\sin^{p+1}(ax+b) + C, \quad \text{if } p \neq -1$$

105. $$\int \sin(ax+b)\cos^p(ax+b)\, dx = -\frac{1}{a(p+1)}\cos^{p+1}(ax+b) + C, \quad \text{if } p \neq -1$$

106. $$\int \sin^2(ax+b)\cos^2(ax+b)\, dx = -\frac{1}{32a}\sin 4(ax+b) + \frac{x}{8} + C$$

107. $$\int \sin^m(ax+b)\cos^n(ax+b)\, dx$$
$$= \frac{-1}{a(m+n)}\Big[\sin^{m-1}(ax+b)\cos^{n+1}(ax+b)$$
$$-(m-1)a\int \sin^{m-2}(ax+b)\cos^n(ax+b)\, dx\Big], \quad \text{if } m > 0, m+n \neq 0$$

See (90) if m is even; (105) if m is odd.

108. $$\int \sin^m(ax+b)\cos^n(ax+b)\, dx$$
$$= \frac{1}{a(m+n)}\Big[\sin^{m+1}(ax+b)\cos^{n-1}(ax+b)$$
$$+(n-1)a\int \sin^{m}(ax+b)\cos^{n-2}(ax+b)\, dx\Big], \quad \text{if } n > 0, m+n \neq 0$$

See (74) if n is even; (104) if n is odd.

109. $$\int \frac{dx}{\sin(ax+b)\cos(ax+b)} = \frac{1}{a}\ln|\tan(ax+b)| + C$$

110. $$\int \frac{dx}{\sin^n(ax+b)\cos(ax+b)} = \frac{-1}{a(n-1)}\frac{1}{\sin^{n-1}(ax+b)} + \int \frac{dx}{\sin^{n-2}(ax+b)\cos(ax+b)}, \quad \text{if } n > 1$$

See (128) if n is even; (109) if n is odd.

111. $$\int \frac{dx}{\sin(ax+b)\cos^n(ax+b)} = \frac{1}{a(n-1)}\frac{1}{\cos^{n-1}(ax+b)} + \int \frac{dx}{\sin(ax+b)\cos^{n-2}(ax+b)}, \quad \text{if } n > 1$$

See (135) if n is even; (109) if n is odd.

112. $$\int \frac{\sin(ax+b)}{\cos^2(ax+b)}\, dx = \frac{1}{a\cos(ax+b)} + C = \frac{1}{a}\sec(ax+b) + C$$

113. $$\int \frac{\sin^2(ax+b)}{\cos(ax+b)}dx = -\frac{1}{a}\left[\sin(ax+b) - \ln\left|\tan\left(\frac{ax+b}{2}+\frac{\pi}{4}\right)\right|\right] + C$$

114. $$\int \frac{\cos(ax+b)}{\sin^2(ax+b)}dx = \frac{-1}{a\sin(ax+b)} + C = -\frac{1}{a}\csc(ax+b) + C$$

115. $$\int \frac{\cos^2(ax+b)}{\sin(ax+b)}dx = \frac{1}{a}\left[\cos(ax+b) + \ln\left|\tan\left(\frac{ax+b}{2}\right)\right|\right] + C$$

116. $$\int \tan x\,dx = -\ln|\cos x| + C = \ln|\sec x| + C$$

117. $$\int \tan(ax+b)\,dx = -\frac{1}{a}\ln|\cos(ax+b)| + C$$

118. $$\int \tan^2(ax+b)\,dx = \frac{1}{a}\tan(ax+b) - x + C$$

119. $$\int \tan^3(ax+b)\,dx = \frac{1}{2a}[\tan^2(ax+b) + 2\ln|\cos(ax+b)|] + C$$

120. $$\int \tan^n(ax+b)\,dx = \frac{1}{a(n-1)}\tan^{n-1}(ax+b) - \int \tan^{n-2}(ax+b)\,dx, \quad n \geq 2$$

See (118) if n is even; (119) if n is odd.

121. $$\int \cot x\,dx = \ln|\sin x| + C = -\ln|\csc x| + C$$

122. $$\int \cot(ax+b)\,dx = \frac{1}{a}\ln|\sin(ax+b)| + C$$

123. $$\int \cot^2(ax+b)\,dx = -\frac{1}{a}\cot(ax+b) - x + C$$

124. $$\int \cot^3(ax+b)\,dx = -\frac{1}{2a}[\cot^2(ax+b) + 2\ln|\sin(ax+b)|] + C$$

125. $$\int \cot^n(ax+b)\,dx = \frac{-1}{a(n-1)}\cot^{n-1}(ax+b) - \int \cot^{n-2}(ax+b)\,dx, \quad n \geq 2$$

See (123) if n is even; (124) if n is odd.

126. $$\int \sec x\,dx = \ln\left|\tan\left(\frac{x}{2}+\frac{\pi}{4}\right)\right| + C = \ln|\sec x + \tan x| + C$$

127. $$\int \sec^2 x\,dx = \tan x + C$$

128. $$\int \sec(ax+b)\,dx = \frac{1}{a}\ln\left|\tan\left(\frac{ax+b}{2}+\frac{\pi}{4}\right)\right| + C$$

129. $\displaystyle\int \sec^2(ax+b)\,dx = \frac{1}{a}\tan(ax+b) + C$

130. $\displaystyle\int \sec^3(ax+b)\,dx = \frac{1}{2a}\left[\sec(ax+b)\tan(ax+b) + \ln\left|\tan\left(\frac{ax+b}{2}\right) + \frac{\pi}{4}\right|\right] + C$

131. $\displaystyle\int \sec^n(ax+b)\,dx = \frac{1}{a(n-1)}\frac{\sin(ax+b)}{\cos^{n-1}(ax+b)} + \frac{n-2}{n-1}\int \sec^{n-2}(ax+b)\,dx, \quad n \geq 2$

See (129) if n is even; (130) if n is odd.

132. $\displaystyle\int \sec x \tan x\,dx = \sec x + C$

133. $\displaystyle\int \csc x\,dx = \ln\left|\tan\left(\frac{x}{2}\right)\right| + C = \ln|\csc x - \cot x| + C$

134. $\displaystyle\int \csc^2 x\,dx = -\cot x + C$

135. $\displaystyle\int \csc(ax+b)\,dx = \frac{1}{a}\ln\left|\tan\left(\frac{ax+b}{2}\right)\right| + C$

136. $\displaystyle\int \csc^2(ax+b)\,dx = -\frac{1}{a}\cot(ax+b) + C$

137. $\displaystyle\int \csc^3(ax+b)\,dx = \frac{1}{2a}\left[-\csc(ax+b)\cot(ax+b) + \ln\left|\tan\left(\frac{ax+b}{2}\right)\right|\right] + C$

138. $\displaystyle\int \csc^n(ax+b)\,dx = \frac{-1}{a(n-1)}\frac{\cos(ax+b)}{\sin^{n-1}(ax+b)} + \frac{n-2}{n-1}\int \csc^{n-2}(ax+b)\,dx, \quad n \geq 2$

See (136) if n is even; (137) if n is odd.

139. $\displaystyle\int \csc x \cot x\,dx = -\csc x + C$

140. $\displaystyle\int \arcsin\frac{x}{a}\,dx = x\arcsin\frac{x}{a} + \sqrt{a^2 - x^2} + C$

141. $\displaystyle\int (\arcsin ax)^2\,dx = x(\arcsin ax)^2 - 2x + \frac{2}{a}\sqrt{1 - a^2x^2}\arcsin ax + C$

142. $\displaystyle\int x \arcsin ax\,dx = \frac{1}{4a^2}\left[(2a^2x^2 - 1)\arcsin ax + ax\sqrt{1 - a^2x^2}\right] + C$

143. $$\int \frac{\textbf{arcsin}\, ax}{x^2} dx = a\, \textbf{ln}\left|\frac{1 - \sqrt{1 - a^2x^2}}{ax}\right| - \frac{\textbf{arcsin}\, ax}{x} + C$$

144. $$\int \textbf{arccos}\frac{x}{a} dx = x\, \textbf{arccos}\frac{x}{a} - \sqrt{a^2 - x^2} + C$$

145. $$\int (\textbf{arccos}\, ax)^2\, dx = x(\textbf{arccos}\, ax)^2 - 2x - \frac{2}{a}\sqrt{1 - a^2x^2}\, \textbf{arccos}\, ax + C$$

146. $$\int \textbf{arctan}\frac{x}{a} dx = x\, \textbf{arctan}\frac{x}{a} - \frac{a}{2}\textbf{ln}(a^2 + x^2) + C$$

147. $$\int \textbf{arccot}\frac{x}{a} dx = x\, \textbf{arccot}\frac{x}{a} + \frac{a}{2}\textbf{ln}(a^2 + x^2) + C$$

148. $$\int \textbf{arcsec}\frac{x}{a} dx = x\, \textbf{arcsec}\frac{x}{a} - a\, \textbf{ln}|x + \sqrt{x^2 - a^2}| + C$$

149. $$\int \textbf{arccsc}\frac{x}{a} dx = x\, \textbf{arccsc}\frac{x}{a} + a\, \textbf{ln}|x + \sqrt{x^2 - a^2}| + C$$

Logarithmic Functions

150. $$\int \textbf{ln}|x| dx = x\, \textbf{ln}|x| - x + C$$

151. $$\int \textbf{log}_a|x| dx = x\, \textbf{log}_a|x| - \frac{x}{\textbf{ln}\, a} + C, \quad \text{if } a \neq 1, a > 0$$

152. $$\int \textbf{ln}|ax + b| dx = \frac{ax + b}{a}\textbf{ln}|ax + b| - x + C$$

153. $$\int (\textbf{ln}|x|)^2\, dx = x(\textbf{ln}|x|)^2 - 2x\, \textbf{ln}|x| + 2x + C$$

154. $$\int (\textbf{ln}|ax + b|)^n\, dx = \frac{ax + b}{a}(\textbf{ln}|ax + b|)^n - n\int (\textbf{ln}|ax + b|)^{n-1}\, dx$$

See (152)

155. $$\int x\, \textbf{ln}|x| dx = \frac{x^2}{2}\textbf{ln}|x| - \frac{x^2}{4} + C$$

156. $$\int \frac{dx}{x\, \textbf{ln}|x|} = \textbf{ln}|\textbf{ln}|x|| + C$$

157. $$\int x^p\, \textbf{ln}|x| dx = x^{p+1}\left[\frac{\textbf{ln}|x|}{p + 1} - \frac{1}{(p + 1)^2}\right] + C, \quad p \neq -1$$

158. $$\int \frac{(\textbf{ln}|x|)^p}{x} dx = \frac{1}{p + 1}(\textbf{ln}|x|)^{p+1} + C, \quad p \neq -1$$

159. $$\int \textbf{sin}(\textbf{ln}|x|)\, dx = \frac{x}{2}[\textbf{sin}(\textbf{ln}|x|) - \textbf{cos}(\textbf{ln}|x|)] + C$$

160. $$\int \textbf{cos}(\textbf{ln}|x|)\, dx = \frac{x}{2}[\textbf{sin}(\textbf{ln}|x|) + \textbf{cos}(\textbf{ln}|x|)] + C$$

Exponential Functions

161. $\displaystyle\int \mathbf{e}^x\,dx = \mathbf{e}^x + C$

162. $\displaystyle\int \mathbf{e}^{ax}\,dx = \frac{1}{a}\mathbf{e}^{ax} + C$

163. $\displaystyle\int x\mathbf{e}^{ax}\,dx = \frac{1}{a^2}\mathbf{e}^{ax}(ax - 1) + C$

164. $\displaystyle\int x^m\mathbf{e}^{ax}\,dx = \frac{1}{a}x^m\mathbf{e}^{ax} - \frac{m}{a}\int x^{m-1}\mathbf{e}^{ax}\,dx, \quad m \geq 2$

See (163)

165. $\displaystyle\int \frac{\mathbf{e}^{ax}}{x}\,dx = \ln|x| + ax + \frac{(ax)^2}{2(2!)} + \frac{(ax)^3}{3(3!)} + \cdots + C$

166. $\displaystyle\int \mathbf{e}^{ax}\sin bx\,dx = \frac{\mathbf{e}^{ax}(a\sin bx - b\cos bx)}{a^2 + b^2} + C$

167. $\displaystyle\int \mathbf{e}^{ax}\cos bx\,dx = \frac{\mathbf{e}^{ax}(a\cos bx + b\sin bx)}{a^2 + b^2} + C$

168. $\displaystyle\int \frac{dx}{1 + \mathbf{e}^x} = x - \ln|1 + \mathbf{e}^x| + C$

169. $\displaystyle\int \frac{dx}{a\mathbf{e}^{px} + b} = \frac{x}{b} - \frac{1}{bp}\ln|a\mathbf{e}^{px} + b| + C, \quad b \neq 0, p \neq 0$

170. $\displaystyle\int \frac{dx}{a\mathbf{e}^{px} + b\mathbf{e}^{-px}} = \frac{1}{p\sqrt{ab}}\arctan\left(\mathbf{e}^{px}\sqrt{\frac{a}{b}}\right) + C, \quad \text{if } ab > 0$

171. $\displaystyle\int \mathbf{e}^{ax}\ln|bx|\,dx = \frac{1}{a}\mathbf{e}^{ax}\ln|bx| - \frac{1}{a}\int \frac{\mathbf{e}^{ax}}{x}\,dx$

See (165).

172. $\displaystyle\int \mathbf{a}^x\,dx = \frac{\mathbf{a}^x}{\ln a} + C, \text{if } a > 0, \quad a \neq 1$

173. $\displaystyle\int \mathbf{a}^{bx}\,dx = \frac{\mathbf{a}^{bx}}{b\ln a} + C, \text{if } a > 0, \quad a \neq 1$

174. $\displaystyle\int x\mathbf{a}^{bx}\,dx = \frac{x\mathbf{a}^{bx}}{b\ln a} - \frac{\mathbf{a}^{bx}}{b^2(\ln a)^2} + C, \quad \text{if } a > 0, a \neq 1$

Hyperbolic Functions

175. $\displaystyle\int \sinh ax\,dx = \frac{1}{a}\cosh ax + C$

176. $\displaystyle\int \sinh^2 ax\,dx = \frac{1}{4a}\sinh 2ax - \frac{1}{2}x + C$

177. $\displaystyle\int \cosh ax\,dx = \frac{1}{a}\sinh ax + C$

178. $\int \cosh^2 ax\, dx = \frac{1}{4a}\sinh 2ax + \frac{1}{2}x + C$

179. $\int \tanh ax\, dx = \frac{1}{a}\ln|\cosh ax| + C$

180. $\int \tanh^2 ax = x - \frac{1}{a}\tanh ax + C$

181. $\int \coth ax\, dx = \frac{1}{a}\ln|\sinh ax| + C$

182. $\int \coth^2 ax = x - \frac{1}{a}\coth ax + C$

183. $\int \operatorname{sech} ax\, dx = \frac{1}{a}\arctan(\sinh ax) + C$

184. $\int \operatorname{sech}^2 ax\, dx = \frac{1}{a}\tanh ax + C$

185. $\int \operatorname{csch} ax\, dx = -\frac{1}{a}\ln|\coth ax + \operatorname{csch} ax| + C = \frac{1}{a}\ln\left|\tanh\frac{ax}{2}\right| + C$

186. $\int \operatorname{csch}^2 ax\, dx = -\frac{1}{a}\coth ax + C$

187. $\int \operatorname{sech} ax \tanh ax\, dx = -\frac{1}{a}\operatorname{sech} ax + C$

188. $\int \operatorname{csch} ax \coth ax\, dx = -\frac{1}{a}\operatorname{csch} ax + C$

B. Definite Integrals

1. $\int_0^\infty e^{-ax^2}\, dx = \frac{1}{2}\left(\frac{\pi}{a}\right)^{1/2}$

2. $\int_0^\infty x^2 e^{-ax^2}\, dx = \frac{1}{4a}\left(\frac{\pi}{a}\right)^{1/2}$

3. $\int_0^\infty x^{2n} e^{-ax^2}\, dx = \frac{1\cdot 3\cdot 5\cdots(2n-1)}{2^{n+1}a^n}\left(\frac{\pi}{a}\right)^{1/2}$

4. $\int_0^\infty x e^{-ax^2}\, dx = \frac{1}{2a}$

5. $\int_0^\infty x^3 e^{-ax^2}\, dx = \frac{1}{2a^2}$

6. $\int_0^\infty x^{2n+1} e^{-ax^2}\, dx = \frac{n!}{2}\left(\frac{1}{a^{n+1}}\right)$

7. $\int_0^\infty x^n e^{-ax}\, dx = \frac{n!}{a^{n+1}}$

8. $$\int_{-\infty}^{+\infty} x^{2n}\mathrm{e}^{-ax^2}\,dx = 2\int_0^{\infty} x^{2n}\mathrm{e}^{-ax^2}\,dx$$

9. $$\int_{-\infty}^{+\infty} x^{2n+1}\mathrm{e}^{-ax^2}\,dx = 0$$

III TRANSFORMATION OF ∇^2 TO SPHERICAL POLAR COORDINATES

The Laplacian operator has its simplest form in Cartesian coordinates:

$$\nabla^2 = \frac{\partial^2}{\partial x^2} + \frac{\partial^2}{\partial y^2} + \frac{\partial^2}{\partial z^2}$$

The transformation and reverse-transformation equations from Cartesian coordinates to spherical polar coordinates are

$$x = r\sin\theta\cos\phi \qquad r = (x^2 + y^2 + z^2)^{1/2}$$

$$y = r\sin\theta\sin\phi \qquad \cos\theta = \frac{z}{(x^2 + y^2 + z^2)^{1/2}}$$

$$z = r\cos\theta \qquad \tan\phi = \frac{y}{x}$$

We must now determine the transformation derivatives:

$$\frac{\partial r}{\partial x} = \frac{1}{2}(x^2 + y^2 + z^2)^{-1/2}(2x) = \frac{x}{r} = \sin\theta\cos\phi$$

Likewise,

$$\frac{\partial r}{\partial y} = \sin\theta\sin\phi \quad \text{and} \quad \frac{\partial r}{\partial z} = \cos\theta$$

A simple way to find $\partial\theta/\partial x$ without having to differentiate the inverse cosine is to differentiate $\cos\theta$ implicitly:

$$-\sin\theta d\theta = -z\left(\frac{1}{2}\right)(x^2 + y^2 + z^2)^{-3/2}(2x)\,dx = -\frac{zx}{r^3}\,dx$$

$$-d\theta = -\frac{\cos\phi\cos\theta}{r}dx$$

$$\frac{\partial\theta}{\partial x} = \frac{\cos\phi\cos\theta}{r}$$

By the same method, we have

$$\frac{\partial\theta}{\partial y} = \frac{\sin\phi\cos\theta}{r} \quad \text{and} \quad \frac{\partial\theta}{\partial z} = \frac{-\sin\theta}{r}$$

To find $\partial\phi/\partial x$, we differentiate $\tan\phi$ directly:

$$\sec^2\phi\, d\phi = -\frac{y}{x^2}dx$$

$$\frac{d\phi}{\cos^2\phi} = -\frac{r\sin\theta\sin\phi}{r^2\sin^2\theta\cos^2\phi}dx$$

$$\frac{\partial\phi}{\partial x} = -\frac{\sin\phi}{r\sin\theta}$$

Likewise,

$$\frac{\partial\phi}{\partial y} = \frac{\cos\phi}{r\sin\theta} \quad \text{and} \quad \frac{\partial\phi}{\partial z} = 0$$

The transformation equations for the first derivative are found with the chain rule:

$$\frac{\partial}{\partial x} = \frac{\partial r}{\partial x}\frac{\partial}{\partial r} + \frac{\partial\theta}{\partial x}\frac{\partial}{\partial\theta} + \frac{\partial\phi}{\partial x}\frac{\partial}{\partial\phi}$$

Substituting the transformation derivatives found earlier into the foregoing equation yields

$$\frac{\partial}{\partial x} = \sin\theta\cos\phi\frac{\partial}{\partial r} + \frac{\cos\phi\cos\theta}{r}\frac{\partial}{\partial\theta} - \frac{\sin\phi}{r\sin\theta}\frac{\partial}{\partial\phi}$$

The other first derivatives are found by a similar method:

$$\frac{\partial}{\partial y} = \frac{\partial r}{\partial y}\frac{\partial}{\partial r} + \frac{\partial\theta}{\partial y}\frac{\partial}{\partial\theta} + \frac{\partial\phi}{\partial y}\frac{\partial}{\partial\phi}$$

$$\frac{\partial}{\partial y} = \sin\theta\sin\phi\frac{\partial}{\partial r} + \frac{\sin\phi\cos\theta}{r}\frac{\partial}{\partial\theta} + \frac{\cos\phi}{r\sin\theta}\frac{\partial}{\partial\phi}$$

$$\frac{\partial}{\partial z} = \frac{\partial r}{\partial z}\frac{\partial}{\partial r} + \frac{\partial\theta}{\partial z}\frac{\partial}{\partial\theta} + \frac{\partial\phi}{\partial z}\frac{\partial}{\partial\phi}$$

$$\frac{\partial}{\partial z} = \cos\theta\frac{\partial}{\partial r} - \frac{\sin\phi}{r}\frac{\partial}{\partial\theta}$$

To find the second-derivative operators, we must apply each of the foregoing operators twice to a function—call it $f(r,\theta,\phi)$—being careful to watch for terms that must be differentiated as products:

$$\frac{\partial}{\partial x}\frac{\partial f}{\partial x} = \sin^2\theta\cos^2\phi\frac{\partial^2 f}{\partial r^2} + \frac{\sin\theta\cos\theta\cos^2\phi}{r}\frac{\partial^2 f}{\partial r\,\partial\theta} - \frac{\sin\theta\cos\theta\cos^2\phi}{r^2}\frac{\partial f}{\partial\theta}$$

$$- \frac{\sin\phi\cos\phi}{r}\frac{\partial^2 f}{\partial r\,\partial\phi} + \frac{\sin\phi\cos\phi}{r^2}\frac{\partial f}{\partial\phi} + \frac{\cos\theta\cos^2\phi\sin\theta}{r}\frac{\partial^2 f}{\partial\theta\,\partial r}$$

$$+ \frac{\cos^2\phi\cos^2\theta}{r}\frac{\partial f}{\partial r} + \frac{\cos^2\phi\cos^2\theta}{r^2}\frac{\partial^2 f}{\partial\theta^2} - \frac{\sin\theta\cos\theta\cos^2\phi}{r^2}\frac{\partial f}{\partial\theta}$$

$$- \frac{\sin\phi\cos\phi\cos\theta}{r^2\sin\theta}\frac{\partial^2 f}{\partial\theta\,\partial\phi} + \frac{\sin\phi\cos\phi\cos^2\theta}{r^2\sin^2\theta}\frac{\partial f}{\partial\phi} - \frac{\sin\phi\cos\phi}{r}\frac{\partial^2 f}{\partial\phi\,\partial r}$$

$$+ \frac{\sin^2\phi}{r}\frac{\partial f}{\partial r} - \frac{\sin\phi\cos\phi\cos\theta}{r^2\sin\theta}\frac{\partial^2 f}{\partial\phi\,\partial r} + \frac{\sin^2\phi\cos\theta}{r^2\sin\theta}\frac{\partial f}{\partial\theta}$$

$$+ \frac{\sin^2\phi}{r^2\sin^2\theta}\frac{\partial^2 f}{\partial\phi^2} + \frac{\sin\phi\cos\phi}{r^2\sin^2\theta}\frac{\partial f}{\partial\phi}$$

$$\frac{\partial}{\partial y}\frac{\partial f}{\partial y} = \sin^2\theta\sin^2\phi\frac{\partial^2 f}{\partial r^2} + \frac{\sin\theta\cos\theta\sin^2\phi}{r}\frac{\partial^2 f}{\partial r\,\partial\theta} - \frac{\sin\theta\cos\theta\sin^2\phi}{r^2}\frac{\partial f}{\partial\theta}$$

$$+ \frac{\sin\phi\cos\phi}{r}\frac{\partial^2 f}{\partial r\,\partial\phi} - \frac{\sin\phi\cos\phi}{r^2}\frac{\partial f}{\partial\phi} + \frac{\cos\theta\sin^2\phi\sin\theta}{r}\frac{\partial^2 f}{\partial\theta\,\partial r}$$

$$+ \frac{\sin^2\phi\cos^2\theta}{r}\frac{\partial f}{\partial r} + \frac{\sin^2\phi\cos^2\theta}{r^2}\frac{\partial^2 f}{\partial\theta^2} - \frac{\sin\theta\cos\theta\sin^2\phi}{r^2}\frac{\partial f}{\partial\theta}$$

$$+ \frac{\sin\phi\cos\phi\cos\theta}{r^2\sin\theta}\frac{\partial^2 f}{\partial\theta\,\partial\phi} - \frac{\sin\phi\cos\phi\cos^2\theta}{r^2\sin^2\theta}\frac{\partial f}{\partial\phi} + \frac{\sin\phi\cos\phi}{r}\frac{\partial^2 f}{\partial\phi\,\partial r}$$

$$+ \frac{\cos^2\phi}{r}\frac{\partial f}{\partial r} + \frac{\sin\phi\cos\phi\cos\theta}{r^2\sin\theta}\frac{\partial^2 f}{\partial\phi\,\partial r} + \frac{\cos^2\phi\cos\theta}{r^2\sin\theta}\frac{\partial f}{\partial\theta}$$

$$+ \frac{\cos^2\phi}{r^2\sin^2\theta}\frac{\partial^2 f}{\partial\phi^2} - \frac{\sin\phi\cos\phi}{r^2\sin^2\theta}\frac{\partial f}{\partial\phi}$$

$$\frac{\partial}{\partial z}\frac{\partial f}{\partial z} = \cos^2\frac{\partial^2 f}{\partial r^2} - \frac{\sin\theta\cos\theta}{r}\frac{\partial^2 f}{\partial r\,\partial\theta} + \frac{\sin\theta\cos\theta}{r^2}\frac{\partial f}{\partial\theta} - \frac{\sin\theta\cos\theta}{r}\frac{\partial^2 f}{\partial r\,\partial\theta}$$

$$+ \frac{\sin^2\theta}{r}\frac{\partial f}{\partial r} + \frac{\sin^2\theta}{r^2}\frac{\partial^2 f}{\partial\theta^2} + \frac{\sin\theta\cos\theta}{r^2}\frac{\partial f}{\partial\theta}$$

Adding the second derivatives gives

$$\frac{\partial^2 f}{\partial x^2} + \frac{\partial^2 f}{\partial y^2} + \frac{\partial^2 f}{\partial z^2} = \frac{\partial^2 f}{\partial r^2} + \frac{2}{r}\frac{\partial f}{\partial r} + \frac{1}{r^2}\frac{\partial^2 f}{\partial\theta^2} + \frac{\cos\theta}{r^2\sin\theta}\frac{\partial f}{\partial\theta} + \frac{1}{r^2\sin^2\theta}\frac{\partial^2 f}{\partial\phi^2}$$

or, as it usually is expressed in operator form,

$$\frac{\partial^2}{\partial x^2} + \frac{\partial^2}{\partial y^2} + \frac{\partial^2}{\partial z^2} = \frac{1}{r^2}\frac{\partial}{\partial r}\left(r^2\frac{\partial}{\partial r}\right) + \frac{1}{r^2\sin\theta}\frac{\partial}{\partial\theta}\left(\sin\theta\frac{\partial}{\partial\theta}\right) + \frac{1}{r^2\sin^2\theta}\frac{\partial^2}{\partial\phi^2}$$

IV STIRLING'S APPROXIMATION

Throughout many areas of physical chemistry, and particularly in the area of statistical mechanics, factorials are used extensively. Recall that N factorial, written $N!$, is defined as

$$N! = (1)(2)(3)\ldots(N-1)(N) \tag{IV-1}$$

where $0! = 1$ and $1! = 1$.

Let us consider the following expression for the natural logarithm of $N!$:

$$\ln N! = \ln 1 + \ln 2 + \ln 3 + \cdots + \ln N$$

$$\ln N! = \sum_{x=1}^{N} \ln x \tag{IV-2}$$

If N is very large, the summation can be replaced by an integral:

$$\ln N! \cong \int_1^N \ln x \, dx \tag{IV-3}$$

Integrating this integral by parts gives

$$\ln N! \cong [x \ln x - x]_1^N = N \ln N - N + 1 \tag{IV-4}$$

If N is much larger than 1, we can write

$$\ln N! \cong N \ln N - N \quad \text{for large } N \tag{IV-5}$$

Equation (IV-5) is known as *Stirling's approximation* and can be used to approximate $\ln N!$ when N is a very large number.

A much more accurate expression for approximating $N!$, particularly for nonintegral values of N, can be found by using gamma functions. This approach, which is beyond the scope of the text, leads to *Stirling's formula*,

$$N! = \sqrt{2\pi N}\left(\frac{N}{e}\right)^N \tag{IV-6}$$

which again is valid as N becomes very large. Taking the logarithm Equation (IV-6) gives

$$\ln N! \cong N \ln N - N + \frac{1}{2}\ln(2\pi N) \quad \text{for large } N \tag{IV-7}$$

Stirling's formula gives better results for smaller values of N.

Let us compare the abilities of Equations (IV-4), (IV-5), and (IV-6) to approximate a relatively small number for $N!$, namely, $60!$. The actual value is 8.32×10^{81}. Equation (IV-4) gives $60! = 1.16 \times 10^{81}$. Equation (IV-5) gives an even poorer approximation: $60! = 4.27 \times 10^{80}$. Equation (IV-6) gives $60! = 8.31 \times 10^{81}$. Clearly, for a value of N as small as 60, Equation (IV-6) gives the best results. As N gets much larger, all three equations work better and better. In chemical statistics, where N usually represents the number of molecules in a macroscopic system (approximately Avogadro's number, 10^{23}), all equations reduce to Equation (IV-5).

V SOLVING A 3 × 3 DETERMINANT

Consider the 3 × 3 determinant

$$\begin{vmatrix} a_{11} & a_{12} & a_{13} \\ a_{21} & a_{22} & a_{23} \\ a_{31} & a_{32} & a_{33} \end{vmatrix}$$

In Chapter 9, we demonstrated how to solve this determinant by using the method of cofactors. An alternative method, which is faster when mastered, uses diagonals. Rather than just list the diagonals (and hope that the student sees the pattern), it may be more useful to observe where the diagonals originate. Following are the six determinants whose six diagonals (circled) are shown (with their signs in front of the determinant):

$$+\begin{vmatrix} \circled{a_{11}} & a_{12} & a_{13} \\ a_{21} & \circled{a_{22}} & a_{23} \\ a_{31} & a_{32} & \circled{a_{33}} \end{vmatrix} \qquad +\begin{vmatrix} a_{11} & a_{12} & \circled{a_{13}} \\ \circled{a_{21}} & a_{22} & a_{23} \\ a_{31} & \circled{a_{32}} & a_{33} \end{vmatrix}$$

$$+\begin{vmatrix} a_{11} & \circled{a_{12}} & a_{13} \\ a_{21} & a_{22} & \circled{a_{23}} \\ \circled{a_{31}} & a_{32} & a_{33} \end{vmatrix} \qquad -\begin{vmatrix} a_{11} & a_{12} & \circled{a_{13}} \\ a_{21} & \circled{a_{22}} & a_{23} \\ \circled{a_{31}} & a_{32} & a_{33} \end{vmatrix}$$

$$-\begin{vmatrix} a_{11} & \circled{a_{12}} & a_{13} \\ \circled{a_{21}} & a_{22} & a_{23} \\ a_{31} & a_{32} & \circled{a_{33}} \end{vmatrix} \qquad -\begin{vmatrix} \circled{a_{11}} & a_{12} & a_{13} \\ a_{21} & a_{22} & \circled{a_{23}} \\ a_{31} & \circled{a_{32}} & a_{33} \end{vmatrix}$$

Therefore,

$$\begin{vmatrix} a_{11} & a_{12} & a_{13} \\ a_{21} & a_{22} & a_{23} \\ a_{31} & a_{32} & a_{33} \end{vmatrix} = a_{11}a_{22}a_{33} + a_{21}a_{32}a_{13} + a_{12}a_{23}a_{31}$$

$$- a_{31}a_{22}a_{13} - a_{21}a_{12}a_{33} - a_{32}a_{23}a_{11}$$

Example

Solve the 3 × 3 determinant, using the method of diagonals:

$$\begin{vmatrix} 1 & 2 & 3 \\ 4 & 5 & 6 \\ 7 & 8 & 9 \end{vmatrix} = (1)(5)(9) + (2)(6)(7) + (4)(8)(3)$$

$$- (7)(5)(3) - (4)(2)(9) - (8)(6)(1)$$

$$= 45 + 84 + 96 - 105 - 72 - 48 = 0$$

VI STATISTICS

The following steps to produce an Excel spreadsheet, together with the accompanying macro, allow one to calculate the mean value, standard deviation, probable error of a single measurement, and probable error of the mean for a collection of measured values:

1. Open Excel.
2. In cell A1, type m(i).
3. In cell B1, type n(i).
4. In cell C3, type mean =.
5. In cell C4, type stand dev =.
6. In cell C5, type prob error =.
7. In cell C6, type error mean =.
8. Go to **TOOLS → MACRO → Visual Basic Editor**. The Project—VBA window will open up. Click on Sheet 1. Press **RETURN**.
9. Type the following macro code exactly as it is shown into the code window:

```
'Macro for Statistical Analysis of Data
Sub Statistics()
DIM dataArray As Variant, rn As Integer, cn As Integer, rnMax As Integer
DIM i As Integer, MVAL(500) As Double, NVAL(500) As Double
DIM MTOT(500) As Double, A As Double, FSUM As Double
DIM RVAL(500) As Double, SVAL(500) As Double, SRSUM As Double
DIM MSUM As Double, STOT(500) As Double, SIGMA As Double,
DIM PROB As Double, PROBMEAN As Double
dataArray = Selection.Value
rnMax = Selection.Rows.Count
For rn = 1 to rnMax
      NVAL(rn) = dataArray(rn,2)
      MVAL(rn) = dataArray(rn,1)
Next rn
FSUM = 0
```

```
For i = 1 to rnMax
   FSUM = FSUM + NVAL(i)
   MTOT(i) = MVAL(i)*NVAL(i)
Next i
MSUM = 0
For i = 1 to rnMax
   MSUM = MSUM + MTOT(i)
Next i
A = MSUM/FSUM
For i = 1 to rnMax
   RVAL(i) = MVAL(i) - A
   SVAL(i) = RVAL(i)*RVAL(i)
   STOT(i) = SVAL(i)*NVAL(i)
Next i
SRSUM = 0
For i = 1 to rnMax
   SRSUM = SRSUM + STOT(i)
Next i
SIGMA = sqr((1/(FSUM - 1))*SRSUM)
PROB = 0.675*SIGMA
PROBMEAN = PROB/sqr(FSUM)
'Output Data
dataArray(1,4) = A
dataArray(2,4) = SIGMA
dataArray(3,4) = PROB
dataArray(4,4) = PROBMEAN
Selection.Value = dataArray
End Sub
```

10. Leave the Code Box by pressing Opt+F11.
11. Save this spreadsheet as "Statistics".

To use the spreadsheet,

1. List the measured values, starting with A3 down to as far as is needed (i.e., A3:A?).
2. List the frequency of each measured value, starting with B3 down to row A?. Each measured value must have a frequency, even if it is only 1.
3. Highlight the cells A3:D?. (This step is important. If you forget to do it, you will get an error message when you call up the macro.)
4. Go to **TOOLS** → **MACRO** → **Macros**. The macro Statistics should be highlighted. If it is not, highlight it. Press **RUN.** The mean value should appear in cell D3, standard deviation in cell D4, probable error of a single measurement in cell D5, and probable error of the mean in cell D6.

Answers

CHAPTER 1

1. First quadrant: $x = +, y = +$; sine +, cosine +, tangent +
   Second quadrant: $x = -, y = +$; sine +, cosine −, tangent −
   Third quadrant: $x = -, y = -$; sine −, cosine −, tangent +
   Fourth quadrant: $x = +, y = -$; sine −, cosine +, tangent −

2. (a) $r = 3.61, \theta = 33.69°$
   (b) $r = 2.45, \theta = 35.26°$
   (c) $r = 4.12, \theta = 75.96°$
   (d) $r = 2.00, \theta = 90°$
   (e) $r = 2.83, \theta = 225°$
   (f) $r = 3.00, \theta = 0°$
   (g) $r = 3.61, \theta = 146.3°$
   (h) $r = 5.00, \theta = 180°$
   (i) $r = 8.94, \theta = 333.4°$

3. (a) $x = 0.860, y = 0.880$
   (b) $x = 0.500, y = 0.866$
   (c) $x = -0.866, y = 1.50$
   (d) $x = -3.29, y = -3.29$
   (e) $x = -2.57, y = 3.06$
   (f) $x = 3.00, y = 0$
   (g) $x = 2.05, y = -1.44$
   (h) $x = 4.24, y = 4.24$

4. (a) $r = 3.74, \theta = 36.70°, \phi = 63.43°$
   (b) $r = 3.74, \theta = 143.3°, \phi = 296.6°$
   (c) $r = 2.24, \theta = 63.49, \phi = 180°$
   (d) $r = 4.12, \theta = 13.86°, \phi = 0°$
   (e) $r = 9.00, \theta = 48.19°, \phi = 63.43°$
   (f) $r = 2.00, \theta = 0°, \phi = 0°$

5. $d\tau = r\, d\theta\, dr\, dz$

6. (a) $|z| = 1.00, \theta = 0°$
   (b) $|z| = 4.00, \theta = 90°$
   (c) $|z| = \sqrt{2}, \theta = 45°$
   (d) $|z| = 5.00, \theta = 306.9°$
   (e) $|z| = 4.24, \theta = 225°$
   (f) $|z| = 7.21, \theta = 123.7°$

7. $x = 1, -0.5 \pm \frac{\sqrt{3}}{2}i$

10. $0, \pm 1, \pm 2, \ldots$

12. $F_{200} = 72$

CHAPTER 2

1. (a) $x = 2$; yes
   (b) $x = 2/3$; yes
   (c) $x = 1 \pm \sqrt{3}i$; no
   (d) $x = -4, +3$; yes
   (e) $x = 2.623, -1.114$; yes
   (f) $x = (2n + 1)\frac{\pi}{2}$; yes
   (g) $\theta = n\pi$; yes
   (h) $(H^+) = 1.00M$; yes
   (i) $x = \pm\sqrt{3}$; yes
   (j) $x = +3 \pm \sqrt{7}i$; no

6. (a) $\frac{-1}{x(x + h)}$
   (c) $3(h + 2x)$

(b) $-\dfrac{(2x + h)}{x^2(x^2 + 2hx + h^2)}$

(d) $\dfrac{-1}{(1 + x)(1 + x + h)}$

7. $k = 6.30 \times 10^{-3}\ \text{min}^{-1}$, $C_0 = 3153$ counts

CHAPTER 3

1. (a) pH = 7.00
   (b) pH = 0.910
   (c) pH = 9.07
   (d) pH = −0.0615
   (e) pH = 1.206
   (f) pH = −1.079
2. (a) $(H^+) = 1.00M$
   (b) $(H^+) = 6.98 \times 10^{-3}M$
   (c) $(H^+) = 0.583M$
   (d) $(H^+) = 1.33 \times 10^{-8}\ M$
   (e) $(H^+) = 3.67 \times 10^{-14}\ M$
   (f) $(H^+) = 1.355M$
3. $(H^+) = 1.11 \times 10^{-7}M$; pH = 6.95
4. $w = -2985$ J
5. $\Delta S = 12.7\ \text{J} \cdot \text{K}^{-1}$
6. $T_2 = 102.6$ K
7. $t = 3302$ years
8. $E = 1.121$ V

CHAPTER 4

1. (a) $9x^2 + 8x - 5$
   (b) $\dfrac{-x}{\sqrt{1 - x^2}}$
   (c) $2x - 9$
   (d) $10 \sec^2 2\theta$
   (e) $x^3 e^{3x}(3x + 4)$
   (f) $A(\cos^2 \theta - \sin^2 \theta)$
   (g) $\dfrac{-x^5 e^{2x}}{\sqrt{1 - e^{2x}}} + 5x^4\sqrt{1 - e^{2x}}$
   (h) $3x^2(1 - e^x)\cos 3x - x^2 e^x \sin 3x + 2x(1 - e^x)\sin 3x$
   (i) $\dfrac{5x^2}{2(1 - 5x)^{3/2}} + \dfrac{2x}{(1 - 5x)^{1/2}}$
   (j) $\dfrac{-e^x}{(1 - e^x)}$
   (k) $-1 - \ln n_i$
   (l) $-3 \ln t\, e^{-3t} + \dfrac{1}{t} e^{-3t}$
   (m) $-\dfrac{A}{t^2} + 1 + \ln t$
   (n) $\dfrac{E^2}{A}\left(2z - \dfrac{27}{8}\right)$
   (o) $-2A\left(\dfrac{N\pi}{L}\right)\sin\left(\dfrac{N\pi x}{L}\right)$
   (p) $\dfrac{\Delta H}{Rt^2}$
   (q) $\dfrac{\Delta G}{Rt^2}$
   (r) $\dfrac{-12A}{r^{13}} + \dfrac{6B}{r^7}$
   (s) $\dfrac{-M}{V^2}$
   (t) $A\left(\dfrac{B}{Rt^2}\right)e^{-B/Rt}$
2. (a) $\left(\dfrac{\partial P}{\partial V}\right)_T = -\dfrac{nRT}{V^2}$
   (b) $\left(\dfrac{\partial P}{\partial V}\right)_T = \dfrac{-nRT}{(V - nb)^2} + \dfrac{2n^2 a}{V^3}$

(c) $\left(\frac{\partial \rho}{\partial T}\right) = \frac{-PM}{RT^2}$

(d) $\left(\frac{\partial H}{\partial T}\right) = b + 2cT - \frac{d}{T^2}$

(e) $\left(\frac{\partial r}{\partial y}\right) = \frac{y}{(x^2 + y^2 + z^2)^{1/2}}$

(f) $\left(\frac{\partial y}{\partial \phi}\right) = -r \sin\theta \sin\phi$

(g) $\left(\frac{\partial^2 S}{\partial P\, \partial T}\right) = \frac{1}{T}\left(\frac{\partial^2 H}{\partial P\, \partial T}\right)$

(h) $\left(\frac{\partial^2 S}{\partial T\, \partial P}\right) = \frac{1}{T}\left(\frac{\partial^2 H}{\partial T\, \partial P}\right) - \frac{1}{T}\left(\frac{\partial V}{\partial T}\right) - \frac{1}{T^2}\left(\frac{\partial H}{\partial P}\right) + \frac{V}{T^2}$

(i) $\sin\theta(\cos^2\phi - \sin^2\phi)$

(j) $\left(\frac{\partial E}{\partial c_B}\right) = \frac{2c_B H_{BB} + 2c_A H_{AB} - E(2c_B + 2c_A S_{AB})}{c_A^2 + c_B^2 + 2c_A c_B S_{AB}}$

(k) $\left(\frac{\partial q}{\partial E_i}\right) = -\frac{1}{kT}\sum e^{-E_i/kT}$

(l) $\left(\frac{\partial q}{\partial T}\right) = \frac{1}{kT^2}\sum E_i e^{-E_i/kT}$

3. (a) slope = 75
(b) slope = 8
(c) slope = 3/4
(d) slope = 3.303
(e) slope = 0
(f) slope = −5
(g) slope = 1.265
(h) slope = 294 $\text{m} \cdot \text{s}^{-1}$
(i) slope = 0.0312
(j) slope = 0.04169
(k) slope = -6.69×10^{-4} M/hr

4. (a) Minimum at $x = 5/6$, $y = -0.0833$
(b) None
(c) $x = (2n + 1)\frac{\pi}{4}$, $n = 0, 1, 2, 3, \ldots$; maxima when n is even, minima when n is odd.
$y = \sin(2n + 1)\frac{\pi}{2}$, $n = 0, 1, 2, 3, \ldots$ Points of inflection $x = n\pi/2$
(d) None
(e) Minimum at $r = 2^{1/6}\sigma$, $U(r) = -e$
(f) Maximum at $\theta = 19.47°$, $\psi = 2.00$
(g) Minimum at $z = \frac{27}{16}$, $E = -z^2\frac{e^2}{a}$
(h) Maximum at $E = \frac{1}{2}kT$
(i) Maximum at $x = \frac{a}{2}$, $P(x) = \frac{2}{a}$
(j) Minimum at $r = \left[\frac{nB}{N_0 A z^2}\right]^{1/n-1}$

5. $\frac{dk}{dT} = \left(\frac{E_a}{RT^2}\right)Ae^{-E_a/RT}$

6. $\frac{d\rho}{dT} = \frac{-PM}{RT^2}$

7. $\left(\frac{\partial P}{\partial T}\right)_V = \frac{nR}{V - nb}$

8. $\left(\frac{\partial P}{\partial V}\right)_T = \frac{-nRT}{(V - nb)^2} + \frac{2n^2a}{V^3}$

9. $\alpha = \frac{R}{RT + bP}$

10. $dV = \frac{-RT}{P^2}dP + \frac{R}{P}dT$

11. Minimum at $r = r_0$

12. $x = a/2$

13. $r = a_0$

CHAPTER 5

1. (a) $y = \frac{5}{4}x^4 + C$

   (b) $y = -\frac{1}{2x^2} + C$

   (c) $y = -\frac{1}{2}\cos 2x + C$

   (d) $y = 9x^4 + 40x^3 + 50x^2 + C$

   (e) $y = 2e^{2x} + C$

   (f) $y = Pv + C$

   (g) $y = RT \ln p + C$

   (h) $y = \frac{1}{2}Mv^2 + C$

   (i) $y = -\frac{Q^2}{r} + C$

   (j) $y = \frac{1}{2\pi W}\sin(2\pi Wt) + C$

2. (a) $y = -\frac{1}{4}e^{-4x} + C$

   (b) $y = \frac{x^3}{3} - A^2x + C$

   (c) $y = \frac{1}{2}\left[x\sqrt{x^2 - A^2} - A^2 \ln\left|x + \sqrt{x^2 - A^2}\right|\right] + C$

   (d) $y = \frac{1}{8}x^8 - \frac{1}{3}x^6 + x^4 + C$

   (e) $y = \frac{x}{2} - \frac{A\sin\left(\frac{2N\pi x}{A}\right)}{4N\pi} + C$

   (f) $y = \frac{x^2}{4} - \frac{xA\sin\left(\frac{2N\pi x}{A}\right)}{4N\pi} - \frac{A^2\cos\left(\frac{2N\pi x}{A}\right)}{8N^2\pi^2} + C$

   (g) $y = \frac{\Delta H}{Rt} + C$

   (h) $y = \frac{e^x(\sin x - \cos x)}{2} + C$

   (i) $y = \frac{t}{2} - \frac{\sin(4\pi Wt)}{8\pi W} + C$

   (j) $y = -\frac{1}{4}\cos^4\phi + C$

   (k) $y = \frac{1}{5}\cos^4\theta\sin\theta + \frac{4}{5}\sin\theta - \frac{4}{15}\sin^3\theta + C$

   (l) $y = \frac{3x}{8} - \frac{3\sin[2(3x + 4)]}{48} - \frac{\sin^3(3x + 4)\cos(3x + 4)}{12} + C$

(m) $y = \frac{1}{2}x^3 \sin 2x + \frac{3}{4}x^2 \cos 2x - \frac{3}{8}\cos 2x - \frac{3x}{4}\sin 2x + C$

(n) $y = \ln\frac{(4 - x)}{(3 - x)} + C$

(o) $y = -\frac{\Delta H}{t} + A \ln t + \frac{B}{2}t + \frac{C}{6}t^2 + C'$

(p) $y = C_P \ln t + C$

(q) $y = \frac{(A - x)^{-n+1}}{n - 1} + C$

(r) $y = -\frac{1}{a^2}e^{-ar}(ar + 1) + C$

(s) $y = -kTe^{-\varepsilon/kT} + C$

(t) $\ln a = -Kt + C$

3. (a) $y = a(T_2 - T_1) + \frac{b}{2}(T_2^2 - T_1^2) + \frac{c}{3}(T_2^3 - T_1^3) + d \ln\frac{T_2}{T_1}$

(b) $y = RT \ln\frac{P_2}{P_1}$

(c) $y = 2\pi$

(d) $y = -\frac{\Delta H}{R}\left(\frac{1}{T_2} - \frac{1}{T_1}\right)$

(e) $y = nRT \ln\frac{V_2 - nb}{V_1 - nb} + n^2a\left(\frac{1}{V_2} - \frac{1}{V_1}\right)$

(f) $y = 1/3$

(g) $y = \frac{a^3}{6} - \frac{a^3}{4n^2\pi^2}$

(h) $y = \frac{1}{4a}\left(\frac{\pi}{a}\right)^{1/2}$

(i) $y = \frac{a_0^2}{4}$

(j) $y = \frac{2k^2T^2}{m^2}$

(k) $y = \frac{1}{a}$

5. (a) $\frac{1}{6}x^3y^2 + C$

(b) $\frac{1}{3}(yx^3 + xy^3) + C$

(c) $\frac{1}{2}y^2(x \ln x - x) + C$

(d) $\left(\frac{1}{2}x^2e^{2x} - \frac{1}{4}e^{2x}(2x - 1)\right)(y \ln y - y)z + C$

(e) 2

(f) $\frac{4}{3}\pi v^3$

(g) $\frac{abc}{h^3}(2\pi mkT)^{3/2}$

7. $\Delta H = aT + \frac{b}{2}T^2 + \frac{c}{3}T^3 + d$

8. $\Delta G = -aT \ln T - \frac{b}{2}T^2 - \frac{c}{6}T^3 + d + eT$

9. 0.198

10. 0.0054

11. $\frac{3}{2}a_0$

CHAPTER 6

1. (a) divergent
   (b) divergent
   (c) convergent; 6
   (d) convergent; 1.178
   (e) convergent; ln 2
   (f) divergent
   (g) convergent; 1.645
   (h) divergent
   (i) divergent
   (j) convergent; 0.6667
2. (a) convergent
   (b) divergent
   (c) test fails
   (d) divergent
   (e) convergent
   (f) convergent
   (g) divergent
   (h) convergent
   (i) divergent
   (j) test fails
3. (a) $-1 < x < 1$
   (b) $-1 < x < 1$
   (c) All values of x
   (d) $-1 < x \leq 1$
   (e) All values of x
   (f) $-1 \leq x \leq 1$
   (g) All values of x
   (h) $0 < x \leq 2$
   (i) $-2 < x < 2$
   (j) $-3 < x < -1$
4. (a) $f(x) = 1 - x + x^2 - x^3 + - \cdots$
   (b) $f(x) = 1 - 2x + 3x^2 - 4x^3 + - \cdots$
   (c) $f(x) = 1 + \frac{1}{2}x - \frac{1}{8}x^2 + \frac{1}{16}x^3 - + \cdots$
   (d) $f(x) = -x - \frac{1}{2}x^2 - \frac{1}{3}x^3 - \cdots$
   (e) $f(x) = 1 - x^2 + \frac{1}{2}x^4 - + \cdots$
   (f) $f(x) = 1 + x \ln a + \frac{(x \ln a)^2}{2!} + \frac{(x \ln a)^3}{3!} + \cdots$
   (g) $f(x) = 1 - \frac{x^2}{2!} + \frac{x^4}{4!} - + \cdots$
   (h) $f(x) = 1 + 3x + \frac{6x^2}{2!} + \frac{6x^3}{3!} = 1 + 3x + 3x^2 + x^3$

10. $g(k) = \frac{2i}{\sqrt{2\pi}k^2}(k\pi \cos k\pi - \sin k\pi)$

11. $g(k) = \frac{\sin kL}{kL}$

14. $g(\omega) = \left[\frac{i \sin(\omega_0 + \omega)(\tau/2)}{2(\omega_0 + \omega)} - \frac{i \sin(\omega_0 - \omega)(\tau/2)}{2(\omega_0 - \omega)}\right]$

CHAPTER 7

1. (a) $y = Ae^{-3x}$
   (b) $y = Ae^{+3x}$
   (c) $y = c_1e^{-x} + c_2xe^{-x}$
   (d) $y = c_1e^{3x} + c_2xe^{3x}$
   (g) $-\frac{1}{(k + k_2)}\ln|ka - (k + k_2)x| = t + C$
   (h) $\phi = Ae^{-ar}$
   (i) $(A) = Ce^{-kt}$
   (j) $\Phi = Ae^{im\phi} + Be^{-im\phi}$

(e) $y = c_1e^{3x} + c_2e^{-3x}$

(f) $y = c_1e^{2ix} + c_2e^{-2ix}$

(k) $y = A\sin\sqrt{\dfrac{k}{m}}x + B\cos\sqrt{\dfrac{k}{m}}x$

(l) $\psi = A\sin\sqrt{\dfrac{8\pi^2 mE}{h^2}}x + B\cos\sqrt{\dfrac{8\pi^2 mE}{h^2}}x$

2. (a) Exact
(b) Exact
(c) Exact
(d) Exact
(e) Exact
(f) Inexact
(g) Exact
(h) Exact
(i) Inexact
(j) Exact
(k) Exact

5. $(\kappa^2 - c^2)a_0 = 0$

7. $f(x) = c_1\sin\dfrac{2\pi x}{\lambda} + c_2\cos\dfrac{2\pi x}{\lambda}$

8. $\lambda = \dfrac{2L}{n}, n = 1, 2, 3, \ldots$

CHAPTER 8

1. (a) $|A| = \sqrt{10}, \theta = 71.57°$
(b) $|A| = 2\sqrt{2}, \theta = 45°$
(c) $|A| = 5, \theta = 306.9°$
(d) $|A| = 2, \theta = 180°$
(e) $|A| = \sqrt{37}, \theta = 260.5°$
(f) $|A| = \sqrt{11}, \theta = 25.24°, \phi = 45°$
(g) $|A| = \sqrt{29}, \theta = 42.03°, \phi = 56.31°$
(h) $|A| = \sqrt{6}, \theta = 114.1°, \phi = 116.6°$
(i) $|A| = \sqrt{11}, \theta = 154.8°, \phi = 255°$
(j) $|A| = \sqrt{2}, \theta = 135°, \phi = 0°$

2. (a) $|C| = 4\sqrt{2}, \theta = 45°$
(b) $|C| = \sqrt{17}, \theta = 75.96°$
(c) $|C| = 3\sqrt{2}, \theta = 45°$
(d) $|C| = 5\sqrt{2}, \theta = 45°, \phi = 53.13°$
(e) $|C| = \sqrt{14}, \theta = 122.3°, \phi = 198.4°$
(f) $|C| = \sqrt{73}, \theta = 134.6°, \phi = 99.46°$

3. (a) 6
(b) 2
(c) −4
(d) 9
(e) −34
(f) −33

4. (a) $|C| = 8, \theta = 180°, \phi$ is undefined
(b) $|C| = 6, \theta = 180°, \phi$ is undefined
(c) $|C| = 13, \theta = 0°, \phi$ is undefined
(d) $|C| = \sqrt{6}, \theta = 65.91°, \phi = 296.6°$
(e) $|C| = 19.52, \theta = 55.70°, \phi = 262.9°$
(f) $|C| = 23.43, \theta = 59.19°, \phi = 153.4°$

8. $L_x = yp_z - zp_y$
$L_y = zp_x - xp_z$
$L_z = xp_y - yp_x$

CHAPTER 9

1. (a) −2
(b) −5
(c) 4
(d) 1
(e) $x^2 - 1$
(f) 1
(g) 18
(h) −91
(i) $x^3 - 2x$
(j) 352
(k) $x^4 - 3x^2b^2 + b^4$

2. (a) $x = \pm 1$
(b) $x = \pm 2$
(c) $x = \pm\sqrt{5}$
(d) $x = 0, \pm\sqrt{3}$
(e) $x = 0, 0, \pm\sqrt{3}$

3. $$\begin{pmatrix} 5 & 1 & 0 & 6 \\ 5 & 3 & -6 & 7 \\ -2 & 3 & 3 & -3 \\ 0 & 8 & 9 & 12 \end{pmatrix}$$

4. (a) $\begin{pmatrix} -6 & 1 \\ 12 & -7 \end{pmatrix}$

(b) $\begin{pmatrix} 4 & -1 \\ 2 & 3 \end{pmatrix}$

(c) $\begin{pmatrix} 12 & 15 & 18 \\ 3 & -1 & -7 \\ 12 & 23 & 38 \end{pmatrix}$

(d) $\begin{pmatrix} 26 & 14 & 38 \\ 10 & 15 & 21 \\ 18 & -14 & 13 \end{pmatrix}$

(e) $\begin{pmatrix} x + 8y + 4z \\ -2x + 3y \\ 5x - y - z \end{pmatrix}$

5. $\begin{pmatrix} 42 & -13 & 11 \\ 68 & -50 & 36 \\ 12 & 6 & -2 \end{pmatrix}$ versus $\begin{pmatrix} 8 & 0 & 34 \\ 4 & -7 & 37 \\ 12 & 29 & -11 \end{pmatrix}$

6. (a) $x = 2, y = 1$
(b) $x = 1, y = 3, z = -4$
(c) $x = 1, y = 1, z = -2, t = -3$
(d) $x = x'\sin\theta - y'\cos\theta$
$y = x'\cos\theta + y'\sin\theta$

11. $\begin{pmatrix} \frac{4}{11} & \frac{1}{11} & \frac{-2}{11} \\ \frac{8}{11} & \frac{2}{11} & \frac{7}{11} \\ \frac{-7}{11} & \frac{1}{11} & \frac{-2}{11} \end{pmatrix}$

12. $\Lambda = \begin{pmatrix} 4 & 0 \\ 0 & -1 \end{pmatrix}$

16. $E_1 = \alpha + 2\beta, E_2 = E_3 = \alpha - \beta$

For E_1: $c_1 = c_2 = c_3 = \dfrac{1}{\sqrt{3}}$

For E_2: $c_1 = \dfrac{1}{\sqrt{2}}, c_2 = -\dfrac{1}{\sqrt{2}}, c_3 = 0$

For E_3: $c_1 = \dfrac{1}{\sqrt{6}} = c_2, c_3 = -\dfrac{2}{\sqrt{6}}$

CHAPTER 10

1. (a) $1 + x + x^2 + x^3 + x^4 + x^5$
(b) $1 - x + x^2 - x^3 + x^4 - x^5$
(c) $\Delta E = E_2 - E_1$
(d) $3x^2y$
(e) $2y^3$
(f) $12x^3y^2$
(g) $8xyz$
(h) $1 + 2x + 3x^2 + 4x^3 + 5x^4$
(i) $x_0! \cdot x_1! \cdot x_2! \cdot x_3! \cdot x_4!$
(j) $\begin{pmatrix} -1 & 0 \\ 0 & -1 \end{pmatrix}\begin{pmatrix} a \\ b \end{pmatrix} = \begin{pmatrix} -a \\ -b \end{pmatrix}$

2. (a) commute
(b) commute
(c) commute
(d) do not commute

7. Eigenvalues $= m\dfrac{h}{2\pi}$

8. Eigenvalues $= \dfrac{n^2h^2}{8ma^2}$

9. Eigenvalues $= a^2$

10. (a) (0.732, 2.732)
(b) (2.121, 3.535)
(c) (4, 3)
(d) (−0.232, 3.598)
(e) (−3.098, 0.634)

12. $$\hat{M}_x = \frac{h}{2\pi i}\left(y\frac{\partial}{\partial z} - z\frac{\partial}{\partial y}\right)$$
$$\hat{M}_y = \frac{h}{2\pi i}\left(z\frac{\partial}{\partial x} - x\frac{\partial}{\partial z}\right)$$
$$\hat{M}_z = \frac{h}{2\pi i}\left(x\frac{\partial}{\partial y} - y\frac{\partial}{\partial x}\right)$$

13. $$\hat{M}_x = -\frac{h}{2\pi i}\left(\sin\phi\frac{\partial}{\partial\theta} + \cot\theta\cos\phi\frac{\partial}{\partial\phi}\right)$$
$$\hat{M}_y = -\frac{h}{2\pi i}\left(-\cos\phi\frac{\partial}{\partial\theta} + \cot\theta\sin\phi\frac{\partial}{\partial\phi}\right)$$
$$\hat{M}_z = \frac{h}{2\pi i}\frac{\partial}{\partial\phi}$$

14. $$\hat{M}^2 = -\frac{h^2}{4\pi^2}\left(\frac{\partial^2}{\partial\theta^2} + \cot\theta\frac{\partial}{\partial\theta} + \frac{1}{\sin^2\theta}\frac{\partial^2}{\partial\phi^2}\right)$$

CHAPTER 11

1. 7427 g NO
2. $a = 1045.6$
   $b = 137.71$
   $c = -42.375$
3. $E_a = 84.13$ kJ
5. $\Delta H = 469.8$ J
6. $\Delta S = 1.599$ J/K
9. (a) $x = -2.926, -0.3341, +1.669$
   (b) $x = -1, 2, 2$
   (c) $x = -0.8622, 1.111, 1.222$
   (d) $x = 1, 1, 1.993, -2.664$
10. $x = -2.070, -0.949, +1.019$

CHAPTER 12

1. 2 points: 1/36
   3 points: 1/18
   4 points: 1/12
   5 points: 1/9
   6 points: 5/36
   7 points: 1/6
   11 points: 1/18
3. $\overline{m} = 5.649$
   $\sigma = \pm 0.018 = \pm 0.02$
   $q = \pm 0.01$
   $Q = \pm 0.001$
4. $V = 0.011 \pm 0.002$ cc
5. $\pm 0.25 = \pm 0.3\ \text{g}\cdot\text{mol}^{-1}$

Index

A

Abscissa, definition of, 1
Absolute value, definition of, 8
Antiderivative, definition of, 57
Antilogarithm, definition of, 26
Areas:
  graphical determination of, 208
  numerical determination of, 175–179
  under curves, 64–67
Arithmetic mean, 86
Associated Laguerre function, 109
Associated Legendre function, 112
Associated spherical harmonics, 112
Auxiliary equations, 101
Average value, 192, 196

B

Base of logarithm, 24
Basis set, 88
Bessel's equation, 122
Boundary conditions, 122, 153

C

Calculus:
  differential, 30–56
  integral, 57–75
Cartesian coordinates, 1–3
  definition of, 1
  differential volume element, 3
  n-dimensional, 2
  three-dimensional, 2
Chain rule, 33, 38
Characteristic equation of a matrix, 141–145
Characteristic of a logarithm, 26
Circles:
  equation of, 19
  graph of, 19
Cofactors, method of, 133–134
Column matrix, 132
Commutation. *See* Commutator.
Commutator, 150
Commutator bracket, 150
Comparison test, 79
Complex conjugate, 8
Complex numbers, 7–9
Complex plane, 7
  definition of, 7
  graph of, 7
Computer, use of, 165–192
  discrete Fourier transforms, 182
  Goal Seek, 179–182
  graphical presentation, 169–175
  macros, 182
  mathematical operations, 166
  method of least squares by, 170
  numerical integration by, 175
    Newton–Cotes method, 177
    trapezoid method, 176
  roots of equations by, 179–182
  spreadsheets, 165–179
  successive approximations, 167
  Trendline, 171
  Visual BASIC, 183
Consecutive reaction, 99
Constant coefficients, linear differential equations with, 100
Constrained maxima and minima, 49–53

Convergence:
of an infinite series, 76
interval of, 82
Convergence and divergence, tests for, 77–81
Convergent series, sums of, 76
Coordinate systems, 1–10
Cartesian:
definition of, 1
differential volume element of, 3
n-dimensional, 2
three-dimensional, 2
curvilinear, 4
cylindrical, 10
plane polar, 3–5
definition of, 4
reverse transformation equations, 5
transformation equations, 4
spherical polar, 5–7
definition of, 5
differential volume element, 6
reverse transformation equations, 6
transformation equations, 6
Cosine, definition of, 19
graph of:
linear coordinates, 20
polar coordinates, 20
Cramer's rule, 141
Cross product, 127
Curves, areas under, 67–70

D

∇^2 operator, 152
transformation to plane polar coordinates, 160–162
transformation to spherical polar coordinates, 225
Derivative, definition of, 30
functions of several variables, 36
functions of single variable, 31
geometric properties of, 45–49
mixed partial second, 37
partial, 36–39
physical significance of. *See* Derivative, geometric properties of.
ratio of infinitesimally small changes, 41–45
relationship to slope of curve, 43
Descartes, René, 1
Determinants, definition of, 133
method of cofactors, 133–134
secular, 142
solving a 3×3, special method of, 229
Differential calculus, 30–56
Differential equations, 95–122
definition of, 95
first order, 97
linear, 95, 100
order of, 95
partial, 117
reduced, 96, 100
second order, constant coefficients, 100
series method of solution, 104
special polynomial solutions:
associated Laguerre polynomials, 110
associated Legendre polynomials, 111
Hermite polynomials, 106
Laguerre polynomials, 108
Legendre polynomials, 111
with constant coefficients:
general method of solution, 100–101
general solutions, 101
imaginary solutions, 101–102
particular solutions, 101
real solutions, 101
Differential volume element, 3
Cartesian coordinates, 3
definition of, 3
spherical polar coordinates, 6
Differentials:
exact, definition of, 113
inexact, definition of, 113
significance of, 115
partial, 40
of a cylinder, 40–41
total, 39–41
Differentiation:
definition of, 31
of functions of several variables, 36–39
of functions of single variable, 31–36
Direct proportion, 13
Discrete Fourier transforms, 182

Divergence:
of infinite series, 77
of a vector, 152
Dot product, 126

E

Eigenfunctions, 153
Eigenvalue equations, 153
Eigenvalue spectrum, 142
Eigenvalues, 142, 153
Eigenvectors, 142
Elements:
of a matrix, 132
of a set, 11
Equations:
auxiliary, 101
differential. *See* Differential equations.
eigenvalue. *See* Eigenvalue equations.
exponential, 18
defining equation, 18
graph of, 18
zero of, 18
first degree, 13
defining equation, 13
graph of, 13
slope of 13
zero of, 14
Gibbs–Helmholtz, 74
Hermite's, 106
indicial, 105
Kirchhoff's, 74
Laguerre's, 108
Legendre's, 111
linear. *See* First degree equations.
logarithmic, 18, 24
definition equation, 18, 24
graph of, 19
polynomial, finding roots of, 179
quadratic. *See* Second degree equations.
recursion, 105
second degree, 14
defining equation, 14
graph of, 15
slope of, 15
zeros of, 15–16
secular, 142
simultaneous. *See* Simultaneous equations.
Error:
average, 198
probable, 197
propagation of, 200–204
random, 196
standard, 197
systematic, 196
Error probability function, 196
Estimated standard deviation, 197
macro for, 230
Exact differentials. *See* Differentials, exact.
Exactness, test for, 113
Excel. *See* Microsoft Excel.
Expansion by cofactors, 133–134
Exponential *e*, definition of, 18
Exponential equations, 18
Exponential functions. *See* Functions, exponential.
Euler's relations for complex exponentials, 8
Euler's test for exactness, 113

F

Factorials, 8, 228
First degree equations, 13
defining equation, 13
graph of, 13
slope of, 13
zero of, 14
Formula, recursion, 105
Fourier series, 85, 182
Fourier integral, 86
Fourier transforms:
continuous, 86
discrete, 182
macro for, 185–186
Free induction decay signal, 189, 192
French curve, 204
Functional dependence of variables, 11–12
Functions:
circular, 19
graph of 19
constrained maxima and minima, 49–53

Functions (*continued*)
  definition of, 11
  differentiation of, 30–56
  even and odd, 83, 88
  exponential, 18
    graph of, 18
    zero of, 18
  graphical representation of, 11–22
  linear, 13
    graph of, 13
    slope of, 13
    zero of, 14
  logarithmic, 18
    graph of, 19
  maximization of, 46–49
  minimization of, 46–49
  quadratic, 15
    graph of, 15
    slope of 15
    zeros of, 15–16
  trigonometric, 20
Functions of state, 115

G

Gaussian distribution, 197
Geometric series, 79
Gibbs–Helmholtz equation, 74
Goal Seek, 179
Gradient operator, 152
Graphical determination of area, 208
Graphs:
  method of least squares, 170
  preparation of using Excel, 169–175

H

Hamiltonian operator, 153
Heat capacity, 42
Heisenberg uncertainty principle, 137
Hermite polynomials, 106
Hermitian operators, 155
Heterogeneous logarithms, 26
Homogeneous logarithms, 26

I

Imaginary numbers, 7
Inexact differentials. *See* Differentials, inexact.
Infinite series, 76–94
  comparison, 79
  convergence of, 76
    tests for, 77–81
  definition of, 76
  divergence of, 77
    tests for, 77–81
  Fourier, 85–90, 182–190
  geometric, 79
  harmonic, 80
  Maclaurin, 83
  power. *See* Power series.
  Taylor, 83
Integral calculus, 57–75
Integrals:
  as an antiderivative, 57–58
  as area under curve, 64–67
  cyclic, 116
  definite:
    definition of, 66
    evaluation of, 66
    table of, 224
  definition of, 58
  double, 79
  Fourier, 86
  geometric interpretation of, 64–71
  indefinite, table of, 211
  line, 67, 115
  partial, 70
  table of, 211
  triple, 70
Integral sign, 58
Integrand, 58
Integrating factor, 99, 116
Integration:
  as a summation, 64
  definition of, 57
  general methods of, 58–59
  graphical, 208
  numerical, 175–179
  partial, successive, 70
  special methods of, 60–64

algebraic substitution, 60
by parts, 62
partial fractions, 62
trigonometric transformation, 61
Interval of convergence, 82

K

Kirchhoff's equation, 74
Kronecker delta, 89, 126

L

Lagrange's method of undetermined multipliers, 49
Laguerre polynomials, 108
Laplacian operator, 152
Least squares determination, 170, 205–207
Legendre polynomials, 111
Line integrals, 67–70, 115
Linear combinations, 96
Linear differential equations. *See* Differential equations, linear.
Linear equations, 13
simultaneous solutions of, 139
Linear functions. *See* Functions, linear.
Linear regression, 170, 205–207
Logarithmic equations, 18
Logarithmic functions. *See* Functions, logarithmic.
Logarithms:
base *e*. *See* Logarithms, natural.
base 10. *See* Logarithms, common.
characteristics of, 26
common, 25
relationship to natural, 27
general properties of, 24–25
power rule, 25
product rule, 24
quotient rule, 24
heterogeneous, 26
homogeneous, 26
mantissa of, 26
Napierian. *See* Logarithms, natural.
natural, 27
relationship to common, 27
significance of, 27–28

M

Maclaurin series, 83
Macros, 183–190
Mantissa, 26
Mathematical set, 11
Matrices:
addition of, 136
general properties of, 132–148
multiplication of, 136
putting in diagonal form, 142–143
Matrix:
characteristic equation of, 141
column, 132
definition of, 132
determinant of, 133
diagonal of, 138
element of, 132
inverse, 138
nonsingular, 138
order of, 132
row, 132
singular, 138
square, 133
unit, 138
Matrix algebra, 135–136
Maxima and minima, 45–49
constrained, 49–53
definition of, 45
Maximization of functions, 45–49
Method of chords, 207
Microsoft Excel, 165–192
Minimization of functions, 45–59
Modulus. *See* Absolute values, definition of.

N

N-factorial (*N*!):
approximation of, 228
definition of, 8
Newton–Cotes method, 177
coefficients, table of, 178
Newton–Raphson method, 179
Normalization, 88, 197
Numerical methods:
integration, 175–179

Numerical methods (*continued*)
- Newton–Cotes method, 177
- trapezoid method, 176
- linear regression, 170–175
- roots of equations:
  - Goal Seek, 179–182
  - Newton–Raphson method, 179

O

Operators:
- addition of, 155
- commutation of, 149–150
- definition of, 149
- differential, 149
- eigenfunctions of, 153
- gradient, 152
- Hamiltonian, 153
- Hermitian, 155
- Laplacian, 152
- rotational, 156–160
- self-adjoint, 155
- transformation, 159
- vector, 151

Ordinate, definition of, 1
Origin, definition of, 1
Orthogonality:
- of functions, 89
- of vectors, 126

Orthonormal set, 89

P

Parabolas, definition of, 15
Partial derivatives. *See* Derivative, partial.
Partial differentials. *See* Differentials, partial.
Partial sums, 76
Phase angle, definition of, 8
Physical constants, table of, 211
Plane polar coordinate, 3–5
- definition of, 4
- graph of, 4
- reverse transformation equations, 5
- transformation equations, 4

Planimeter, 209
Point of inflection:
- definition of, 47
- test for, 47

Power series:
- definition of, 81
- Fourier, 85–90
- interval of convergence, 82
- Maclaurin, 83
- Taylor, 84

Power spectrum, 183
Probability, definition of, 193
Probability aggregate, 197
Probable error, 197
- macro for, 230
- of a single measurement, 198
- of the mean, 198

Propagation of errors, 200–204
Pycnometer, 202

Q

Quadratic equations. *See* Second degree equations.
Quadratic formula, 16
Quadratic functions. *See* Functions, quadratic.
Quadrature, 175

R

Radioactive decay, 23
Random error, 196
Ratio test, 80
Real numbers, 7
Rectangular coordinates. *See* Cartesian coordinates.
Recursion equation, 105
Recursion formula, 105
Reduced differential equation, 96
Residuals, 196
Reversibility, 43, 68
Reversible processes. *See* Reversibility.
Riemann sums, 208
Right hand rule, 127

Roots. *See under* Zeros.
Rotational operators, 156–160
relationship to symmetry, 160
Round space. *See* Space, round.
Row matrix, 132

S

Scalar product, 126
Scalars, definition of, 123
Second-degree equations, 14–17
defining equation, 14
graph of, 15
slope of, 15
zero of, 15–16
Secular determinant, 142
Secular equations, 142
Self-adjoint operators, 155
Separation of variables, 97, 118
Series. *See* Infinite series; Power series.
Series expansion of functions. *See* Fourier series; Maclaurin series; Taylor series.
Sets. *See* Mathematical sets.
Simpson's rule, 178
Simultaneous equations, solutions of, 139
Sine, definition of, 19
Slope:
definition of, 13, 15, 30
determination using differential calculus, 41–45
of linear functions, 13
of quadratic functions, 15
variation of, graph of, 16
Space:
rectangular, 1
round, 4
Spectrum, power, 183
Spherical polar coordinates, 5–7
definition of, 5
differential volume element, 6
reverse transformation equations, 6
transformation equations, 6
Spreadsheets:
for discrete Fourier transforms, 182–190
for graphical presentation, 169–175
for least squares calculations, 170–175
for numerical integration, 175–179
for roots of polynomial equations, 179–182
Square matrix, 133
Standard deviation. *See* Standard error.
Standard error, 197
State functions, 115–116
Stirling's approximation, 228
Straight lines, 13–14
equations for, 13
method of least squares. *See* Linear regression.
Successive approximations, 167
Successive partial integration, 70
Sums, partial, 76
Systematic errors, 196

T

Tangent, definition of, 19
determination of:
using differential calculus, 41–45
graphical method, 207
Taylor series, 84
Total differentials. *See* Differentials, total.
Transformation of ∇^2:
to plane polar coordinates, 160–162
to spherical polar coordinates, 225
Trapezoid method of integration, 176
Trendline, 171
Trigonometric identities, 158

U

Uncertainty principle, Heisenberg, 137
Undetermined multipliers, 50
definition of, 50
Unit circle, 19
Unit matrix, 138
Unit vectors, 124
cross product of, 127
dot product of, 126

V

Variables:
  fractional change, 27–28
  fractional dependence of, 27–28
  separation of, 97, 118
Vector operators, 151–153
Vector product, 127
Vectors:
  absolute value of, 124
  addition of, 123–124
  applications of, 128
  definition of, 123
  magnitude of, 124
  matrix representation of, 142
  scalar multiplication of, 126
Visual BASIC, 183
Volume elements. *See* Differential volume elements.
Volumes, by triple integration, 70–71

W

Wave functions, definite energy, 88
Work:
  area under curve, 64
  pressure-volume, 64, 65, 116

X

x-axis, definition of, 1

Y

y-axis, definition of, 1
y-intercept, 13

Z

Zeros:
  of exponential functions, 18
  of linear functions, 14
  of quadratic functions, 15–16
  roots of polynomial equations, 179–182